DEDICATION

To organizations that want successful audiovisual systems and to the professionals who know how to make it happen.

The best way possible.

AUDIOVISUAL BEST PRACTICES

**THE DESIGN AND
INTEGRATION PROCESS
FOR THE AV
AND CONSTRUCTION
INDUSTRIES**

International Communications
Industries Association, Inc.®

Addressing AV process questions of architects, building and construction trade personnel, consultants, contractors, developers, engineers, facility owners, project managers and AV professionals.

First edition.

Published by the International
Communications Industries
Association, Inc.® (ICIA®)

11242 Waples Mill Road
Fairfax, VA 22030, U.S.A.

www.infocomm.org

Copyright 2005 by the International
Communications Industries Association, Inc.®

Authors:
Timothy W. Cape, CTS-D, Principal
Consultant, Technitect, LLC
Jim Smith, CTS, Director of Strategic
Initiatives, HB Communications, Inc.

Editors:
Susan J. Higginbotham
Taly Walsh, Senior Vice President,
Marketing & Membership, InfoComm
International

Book Design:
Holli Rathman

Printed in the United States of America
Library of Congress Control Number
2005927586

ISBN: 0-939718-20-0

All rights reserved. No part of this book
may be reproduced in any form or by any
electronic or mechanical means, including
photocopying and information storage and
retrieval systems, without written permission
from the publisher.

While every precaution has been taken in
the preparation of this book, the publisher
assumes no responsibility for errors or
omissions. Neither is any liability assumed
for damages resulting from the use of the
information contained herein.

WHAT THIS BOOK MEANS TO INFOCOMM INTERNATIONAL EXECUTIVE DIRECTOR, RANDAL A. LEMKE, PH.D.

As I was reading through the first draft of *Audiovisual Best Practices*, I paused at a paragraph that reads, *"The greatest differentiation between pro-AV and other trades is that pro-AV involves the full-blown creation of communications environments. Consequently, AV professionals are concerned with more than just the cable pathway and the electronics at the end of the cable; they are concerned with making it possible for the customer to communicate."*

This statement accurately describes our industry. AV professionals, who have found their life's work in this field, are increasingly more concerned with the bigger picture in systems integration. They and their collaborators from other disciplines need to know, from the outset, how the entire building construction or renovation will affect the AV system, and vice versa. This new perspective means that the AV professional, architect, engineer, facility owner, project manager and others in the process interface routinely with one another. Successful AV installation and integration is no longer dependent upon a small group of technically savvy individuals (although they are vital to the process), but rather it requires a strong relationship among the members of a more disparate team. To be effective, this team must know as much as possible about AV projects and how they are accomplished.

This book, then, is our opportunity to disseminate information about the business of pro-AV. It is a unique publication that reveals the best of AV — from the standpoint of the industry, its roots, its professionals, its processes and practices and, importantly, its customers.

To see this book published during my tenure as Executive Director is right up there with being a part of an industry that is emerging as a powerhouse and helping the association make valuable contributions to its membership and the industry as a whole.

To all those who worked tirelessly on this book, as well as those who conceived of it and made it happen, I offer simply this statement: "You set a high goal when you conceived of this book, and clearly you have reached it."

Randal A. Lemke, Ph.D.
Executive Director
International Communications Industries Association, Inc.

The InfoComm Best Practices Committee

Committee Chair
Mark Valenti
    The Sextant Group

Co-Authors
Tim Cape, CTS-D
    Technitect
Jim Smith, CTS
    HB Communications
    SAVVI Council Chair

SAVVI
George Bing
    HB Communications
Spencer Bullins, CTS
    AMX
Tom Peters, CTS
    Integrated Media Systems
    2005 InfoComm President
Jim Smith, CTS (co-author)
    HB Communications
    SAVVI Council Chair
Byron Tarry, CTS
    AVW-TELAV Audio Visual Solutions

ICAT
Tim Cape, CTS-D (co-author)
    Technitect
Kris Kuipers
    Newcomb & Boyd
Jeff Loether
    Electro-Media Design
Mark Valenti
    The Sextant Group
    AVBP Committee Chair
Brad Weber
    muse, inc.

Technology Managers/End Users Council
John Pfleiderer, CTS-D
    Cornell University
    Technology Manager/End-Users Council Chair

Reviewers
Peter Gross, AIA
    Kohn Pederson Fox Associates PC
    Architects and Planning Consultants
Stephen Newbold, AIA, RIBA
    Gensler
Dave Labuskes, RCDD/NTS/OSP, CSI
    RTKL Associates, Inc.
John Pfleiderer, CTS-D
    Cornell University
Scott Walker
    Waveguide Consulting, Inc.
    2005 InfoComm Chairman
Brad Weber
    muse, inc.

Editors
Susan Higginbotham
    Higginbotham Associates
Taly Walsh
    InfoComm

ACKNOWLEDGMENTS

Audiovisual Best Practices: The Design and Integration Process for the AV and Construction Industries is the result of an extraordinary combined effort of three primary groups among the membership of the International Communications Industries Association, Inc.® (ICIA®). They are the AV systems integrators, the independent consultants and the technology managers/end-users.

Through their councils — the Sound, AudioVisual and Video Integrators (SAVVI), the Independent Consultants in Audiovisual Technology (ICAT) and the Technology Managers/End-Users — members volunteered their time and commitment in forming the AV Best Practices Subcommittee, chaired by Mark Valenti of The Sextant Group.

The result was a committee of eleven that gathered and determined the scope and goals of the project, the intended audiences, and the elements that were necessary for the completion of a useful and valuable guide to the AV process.

The collaboration of integrators and consultants was a significant accomplishment that merits a word of recognition. While often serving as team members on projects, independent consultants and systems integrators sometimes find themselves at odds philosophically. They also often compete for business. Yet, when the two groups decided to work together on this project for the good of the AV industry, they rolled up their sleeves, put aside differences, agreed that sometimes it was OK to disagree, and got to work.

The significant body of work was developed by the co-authors, Tim Cape, CTS-D, of Technitect, and Jim Smith, CTS, of HB Communications.

Our heartfelt thanks go also to Bradley P. Weber, P.E., of muse, inc., who was a major contributor and peer reviewer, as well as to architects, design consultants, integrators and technology managers who provided insight and in-depth reviews: Peter Gross, AIA, of Kohn Pederson Fox Associates PC; David Labuskes, RCDD/NTS/OSP, CSI, of RTKL Associates, Inc.; Jeff Loether of Electro-Media Design; Stephen Newbold, AIA, RIBA, of Gensler; John Pfleiderer, MA, CTS-D, of Cornell University; Byron Tarry, CTS, of AVW-TELAV Audio Visual Solutions; and Scott Walker, CTS-D, of Waveguide Consulting, Inc.

Susan J. Higginbotham of Higginbotham Associates and Taly Walsh of InfoComm edited the publication; creative design and layout of the publication were produced by Holli Rathman.

The publisher, InfoComm, is the association for the professional audiovisual industry. InfoComm represents the entire distribution chain worldwide, including AV technology manufacturers, distributors, dealers, systems integrators, independent representatives, rental and staging companies, independent consultants, independent programmers, production companies, presentations professionals and technology managers/end-users.

Through its Board of Governors, its councils and committees, and members at large, InfoComm International has advanced numerous industry initiatives, including an AV industry awareness campaign, online and classroom education, individual and company certification at general and specialized levels for design and installation, workforce development, awards programs, online information resources, and InfoComm and Integrated Systems tradeshows for Europe, China and Asia.

The 2005 *Audiovisual Best Practices* guide, and its companion, *The Basics of Audio and Video Systems Design*, are two key initiatives to emerge from the association. InfoComm gratefully acknowledges the important contributions of its membership toward making these valuable industry references a reality.

PREFACE

The realization that AV was becoming a significant industry was reinforced by the issuance in the late eighties of InfoComm International's milestone publication, *"The Basics of Audio and Visual Systems Design."*[1] This landmark book, updated and re-published in 2004, delineated the complex technical aspects of professional AV design.

The AV industry, however, has become far more than simply technical design and application. Because of its pivotal role in delivery of modern information communications, it is an integral part of virtually all major construction and redevelopment projects, as well as the production of live events. As such, the industry has experienced record growth. According to a 2004 study, the AV industry represents a total market size of close to $19 billion in North America alone.[2] In actuality, the worldwide statistics may exceed triple that figure.

Because of the industry's increasing significance, three leading member councils at ICIA[3] recognized that the time had come to produce another industry publication that would go beyond the basics. This new book would help industry professionals, as well as those with whom they work, better understand the process of managing complex information communications projects in buildings and facilities.

To accomplish this goal, the content of *Audiovisual Best Practices* needed to cover significant territory. It needed to present an overview of the industry, explore the inner workings of AV projects with start-to-finish process descriptions, and conclude with an assessment of what the

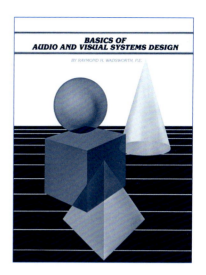

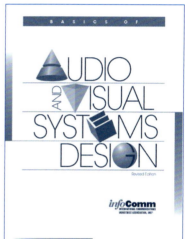

Members of the AV Best Practices Committee during its meeting in October 2004. Pictured, l. to r., Jim Smith, CTS, Kris Kuipers, Tim Cape, CTS-D, Byron Tarry, CTS, Brad Weber, Mark Valenti, Jeff Loether, John Pfleiderer, CTS-D. Not pictured: George Bing, Spencer Bullins, CTS, and Tom Peters, CTS.

---

[1] Raymond Wadsworth, *The Basics of Audio and Visual Systems Design* (Fairfax, Virginia: InfoComm. 1983). The book was the first to contain the diverse set of physics, electronics, ergonomics and design that makes up the audiovisual system. Many AV professionals credit this book for advancement in their careers.
[2] Survey conducted May 2004 for InfoComm/ICIF by Acclaro: Market Intelligence Special Report: 2004 AV Market Definition and Strategy Study.
[3] The Sound, AudioVisual and Video Integrators (SAVVI), the Independent Consultants in Audiovisual Technology (ICAT) and the Technology Managers/End-Users Councils

The AVBP Committee spent a day in October 2004 wrestling with the goal, scope, audiences and outline of the guide.

future holds for the industry.

While the development of this book has been professionally rewarding for all those involved, it was a complicated venture from the outset. The job of producing a single volume that encompassed all aspects of the delivery of AV installations — from project design to development and from installation to systems operation and training — was no easy task.

In addition, the book had to appeal to a multi-faceted audience that existed within and beyond the AV professional's arena. These diverse groups include architects, mechanical engineers, electrical engineers, structural engineers, general contractors, electrical contractors, facility managers and owners, and many others who perform vital roles in the process. Finally, the publication needed to have relevance to the end-users — critical players on the team because they need to "own the project" and will reap the rewards of its success.

Creating a book of this nature required a first-of-its-kind industry collaborative effort. Professionals from every aspect of the industry (in many cases, those who would be competitors outside of the conference room) sat together over a two-day period and planned this remarkable volume. Integrators, designers, and end-users — typically coming to the project from differing perspectives — made a commitment that transcended their usual business interests. They joined forces, putting the best interest of the AV industry first. This book is the result of that effort.

It should be noted that the processes and practices described in the guide are primarily derived from North American experience; however, with InfoComm International's international focus, attempts are made to allow for variation in other parts of the world, while still offering a foundation and starting point to serve as a guideline for future adaptations.

Some aspects of AV solutions, including live events, video production and presentations, are mentioned in passing in *Audiovisual Best Practices: the Design and Integration Process for the AV and Construction Industries.* While these are important aspects of the AV industry, this publication's scope is focused on fixed installations.

TABLE OF CONTENTS

From InfoComm International Executive Director i

Acknowledgments ii

Preface iii

Introduction 1
    Looking Back on the Industry
    Defining the AV Industry
    The Demand for AV Professionals

Overview of the Design and Construction Process 19

Chapter 1 — Understanding the Project Team 23
    An Overview of the Teams
    The Owner Team
    The Design Team
    The Installation Team
    The Management Team
    Roles within Organizations
    Others in the Design and Construction Process

Chapter 2 — Understanding the Process. 39
    Assessing Project Characteristics
    Funding AV Systems Installation and Support
    Legal and Policy Issues
    Owner and End-User Expertise
    Examining the Processes

Chapter 3 — Selecting and Contracting the Project Team. 55
    Considerations in Building the Project Team
    AV Provider Selection Strategies
    Structuring the RFQ and the RFP
    Scoping Strategies for Contracting
    AV Integration Contract Issues
    Contracting for AV Software

Chapter 4 — The Program Phase 79
    Step One: Review the Existing Documentation and Facilities
    Step Two: Benchmark Comparable Facilities
    Step Three: Conduct the Program Meetings
    Step Four: Write the Program Report
    Step Five: Distribute the Report
    Step Six: Approve the Program Report as a Basis for Systems Design

Chapter 5 — The Design Phase 103
    Step One: Start the Design Phase with a Kick-Off Meeting
    Step Two: Establish the Infrastructure
    Step Three: Developing the AV System Design Package
    Step Four: Making the Most of Review

Chapter 6 — The Construction Phase 129
    Project Coordination, Scheduling and Sequencing
    Project Communications and Contract Changes
    Step One: The Construction Kick-Off Meeting
    Step Two: Preparing for Submittals
    Step Three: Procuring the AV Equipment
    Step Four: Preparing the Site
    Step Five: Pre-Assembling and Testing the AV System
    Step Six: Site Installation
    Step Seven: Finalizing the Documentation

CHECKLISTS

AV Provider Scope of Work . . 76
RFQ 78

AV Task Parameters for
Development of the
AV Program 100
AV Needs Analysis/Program
Meeting Agenda Sample . . 101
Benchmarking 102

Design Phase Infrastructure
Coordination 126
Outline of Typical RFP/Design
Package Components 127

AV Construction Kick-Off
Meeting Agenda 148
Site Readiness and Security 150
Shop and As-Built Drawing
Components 151

Substantial Completion 170
Final AV System Record Documentation Package ... 171

Owner Responsibility 188

Chapter 7 — System Commissioning and Training 153
The Basics of System Commissioning
Step One: Perform Preliminary Tests on the Completed System Installation
Step Two: Generate the Punch List
Step Three: Establish Substantial Completion
Step Four: Inspect, Test and Align
Step Five: Train the Users
Step Six: Sign-Off and Start the Warranty Period

Conclusion and Afterword 173

Appendices
I. Contract Forms 181
II. Program Phase 193
III. Construction Phase 197
IV. Commissioning and Training 198
V. Resources and Bibliography 202
VI. Glossary 203
VII. Index 208
VIII. Authors' Biographies

Figures and Charts
Figure 1. AV Market Breakdown 2
Figure 2. AV Industry Growth Rate 3
Figure 3. AV Channels – Revenues 10
Figure 4. AV Process Overview 20
Figure 5. Project Team Overview 24
Figure 6. Large Construction Project: Typical Schedule 44
Figure 7. Smaller Construction Project: Typical Schedule 45
Figure 8. Typical Consultant-Led Design-Bid-Build Contract Structure 47
Figure 9. Typical Integrator-Led Design-Build Contract Structure 48
Figure 10. Method Selection Chart 50
Figure 11. Typical Consultant-Led Design-Build Contract Structure 51
Figure 12. Typical OFE/Integrator Installed Contract Structure 52
Figure 13. Typical OFE/Owner Installed Contract Structure 54
Figure 14. Form SF330 62
Figure 15. The Needs Analysis Pyramid 81
Figure 16. Translating the Needs into a Design 82
Figure 17. The Program Phase Process 83
Figure 18. The AV Design Process 105
Figure 19. The Sightline Study 108
Figure 20. Matching up Contract Deliverables 121
Figure 21. The Typical AV Integration Process 137
Figure 22. Relative Schedules during Construction 140
Figure 23. The Commissioning, Training and Sign-Off Process 147
Figure 24. Sample Custom End-User Operation Guides 164
Figure 25. Degree of Opportunity for Revenue Growth 176
Figure 26. Most Attractive Markets 177
Figure 27. Most Important End-User Markets 178

Photos and graphics for *Audiovisual Best Practices* were provided courtesy of:
AMX Corporation; Audio Visual Innovations, Inc.; Brett Sneed/Swiderski Electronics Inc.; Cinema Park Network, Ltd.; Crestron Electronics, Inc.; Ewing Consulting; Gray Bow Communications; the International Communications Industries Association, Inc.® (ICIA®); HB Communications; Konover Construction Corporation; Marigan O'Malley-Posada / RTKL Associates Inc.; MediaMethods; Media Systems Design Group; NOR-COM, Inc.; Smith Audio Visual, Inc.; Swanson Rink, Inc.; Technical Innovation, LLC; Technitect, LLC; United Visual, a Chicago based audiovisual system integrator and AV dealer; and Waveguide Consulting, Inc.

INTRODUCTION

INTRODUCTION

The professional audiovisual (pro-AV) industry has come into its own.

Never before in its history has professional audiovisual system design and integration been such a vital part of the construction industry. In 2004 alone, the industry reported revenues of nearly $19 billion in North America, with a projected growth rate of 9.6% over the next five years; the worldwide statistic is potentially more than triple that figure. The AV industry has undoubtedly

Figure 1. AV Market Breakdown
$18.9 billion total market size for the year 2003 in the US and Canada

Products = $11.5 Billion
(Figures shown in millions)

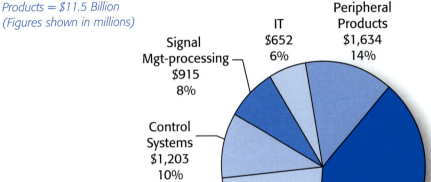

IT	Peripheral	Display	Audio/Video
Wireless Connectivity & Software $380	Furniture $402	Displays $1,115	Acquistion & Delivery $1,025
	Cables/Connect $425	Projectors $2,400	Conferencing $534
Streaming Media & Webcasting $272	Mounting $422	Screens/Shades $650	Sound Reinforcement $942
	Lighting $385	White Boards $390	

According to the 2004 Market Definition and Strategy Study conducted by Acclaro Partners for InfoComm, the total North American AV market size of $18.9 billion was largely above what had been anticipated. The convergence of AV and IT as well as new applications for AV products are key factors driving the market. Growth in the product component of the total market is driven primarily by display systems. That product category has witnessed acceleration in innovation and new technologies, particularly in plasma and flat screens, as well as in projectors.

Source: InfoComm 2004 AV Market Definition and Strategy Study

become a major player in the global economy.

The AV industry has evolved dramatically. AV technology enhances communication everywhere today — whether in the boardroom, the classroom or in public spaces, such as airports, museums and retail centers. The convergence of AV and IT, emerging applications for AV products, as well as the demand driven by home markets are key factors driving the professional AV market. These trends promise a continuing significant impact within and outside of the AV industry.

AREAS OF GROWTH

The increased usage of select audiovisual technologies in key market segments is behind the growth of the industry. Beyond its top markets (business/corporate, colleges/universities, and government/military), AV technology is making significant gains in other arenas. These growth areas include retail establishments, sports facilities, museum exhibits, meeting rooms, convention sites, healthcare facilities, and, interestingly, places of worship. So much growth is expected that, as of 2004, 75% of the audiovisual companies responding to an InfoComm International survey were confident enough to forecast a revenue and profitability increase over the next few years.

From videoconferencing and streaming media to digital signage[4] and plasma displays, AV technology is virtually everywhere. Innovative AV solutions are appearing in all market segments, reflecting the increase in mainstream awareness of AV technologies. The general population is seeing technology as a necessary component in homes[5], businesses, and educational facilities. In fact, AV technologies are being incorporated anywhere there is a need or desire for state-of-the-art communication or visualization. The AV industry has

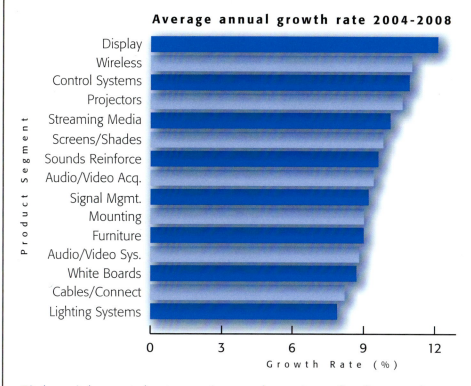

Displays, wireless, control systems, projectors and streaming media will outpace the overall industry growth rate of 9.6%, spurred on by technological innovations and the addition of customer-driven features such as wireless and interoperability. Streaming media, webcasting, wireless and software are all relative newcomers to the AV space. Technologies are being introduced rapidly as a result of AV and IT convergence. Currently, these segments have low penetration rates; however, they are expected to grow significantly as products continue to migrate to web-enable technologies. Cables, connectors, furniture and lighting systems are viewed as "must haves" but only represent a fringe contribution to overall revenue.

Source: InfoComm 2004 AV Market Definition and Strategy Study

---

[4] A positive trend is the increased use of digital signage, particularly in the retail sector, for delivering messages, selling products or simply providing information. The increased demand (primarily from business and universities) has led to increased competition, which, in turn, has lowered costs. The cycle continues, and the industry grows.

[5] Both here and abroad, the increasingly sophisticated residential market is growing. Nearly 25% of dealers in North America and nearly 30% in Europe are currently in residential sales. Evidence of the increasing crossover between commercial and residential applications is the launch of a Residential Pavilion at InfoComm's InfoComm 2005, the largest tradeshow for the professional audiovisual industry worldwide. The pavilion features home theater systems, high-definition displays, residential audio, lighting, electronics and security systems, satellite equipment, and home automation systems.

energized itself to meet these needs and, by all predictions, has begun a long run of strong growth.

Looking Back on the Industry

The industry has come a long way. Not long ago, it was known primarily as the source for overhead projectors, projection screens and basic audio equipment. Today, the AV business

consists of designing and installing sophisticated multimedia, integrated systems that are integral

A Retrospective of Audiovisual Progress

1849 — Lantern slides introduced for viewing photos in small group format.

1887 — Hertz sends and receives radio waves.

1889 — First motion picture system developed.

1893 — First full-length movie, The Great Train Robbery, released.

1894 — Thomas Edison introduces the Kinetoscope for viewing a continuous film reel.

1897 — German scientist, Karl Braun, invents the cathode ray tube oscilloscope (CRT), the basis of television.

1919 — Hoxie develops a means of recording sound on motion picture film.

1924 — GE patents the modern moving coil, direct radiator and loudspeaker sold to consumers under the name "Radiola."

1927 — Philo Farnsworth is first inventor to transmit a television image.

1936 — AEG Co. in Germany invents tape recorder using plastic tape coated with magnetic material.

1938 — Marzocci files patent application for rotary head audio recorder.

1939 — First television sets offered for sale in U.S.

1941 — Hi-Fi recording begins in Germany during World War II.

1953 — Wireless microphone demonstrated.

1950 — First Sony tape recorder marketed in Japan.

1953 — Color TV broadcasting begins.

1961 — KODAK introduces first in its successful line of CAROUSEL projectors.

to a building's environment, as well as the creation of live events for sports, entertainment and corporate presentations.

As the decades passed, audio and video technology evolved to become the basis for the information communications age. Advancements have improved the quality of life from providing photo-keepsakes for Civil War widows to a ground-breaking historical moving picture retrospective, "The Birth of a Nation" and from job training films made during the Depression to contemporary wired classrooms for college students. As events unfolded, technologies advanced, and the world would never be the same.[6]

In response to a rapidly changing world of technology, the Cleveland, Ohio School System felt the need to plan a unique conference. It was now 1937 and the meeting was planned to help the schools identify, better understand and organize their visual classroom materials.

Manufacturers, dealers and producers were in attendance to show their wares. At this meeting, the idea for a national audiovisual organization was formed. Talk turned to action, and the National Association of Visual Education Dealers (NAVED) became a reality.

With the advent of World War II, the industry, with the help of NAVED, geared up to train millions of men and women to conduct the business of war. The U.S. government needed both hardware and software in the form of slides, projectors, and transparencies to train its under-prepared military and civilian population.

When the war was over, weary soldiers returned home, anxious for a life of peace and prosperity. The audiovisual industry made a smooth transition from a half-decade of war service to the education demands of the post-war years. NAVED, revitalized with the energy and experience of post-military personnel, re-emerged stronger than ever.

Three years after the war, NAVED and the Non-Theatrical Film Association merged to form the National

[6] A significant resource for this section was the 1980 slide presentation, "Audio-Visuals: A History," created by the Past Presidents of the National Audio-Visual Association. Written and produced by Terrence M. Bolls, Roger Troup and Harry McGee

NAVA CREED

Be on the forefront

of every audio-visual

development.

Guide our members

to new and

productive uses

of audio-visuals

and to new markets.

Develop new methods

of training.

Set the audio-visual

standard of

excellence for

the world.

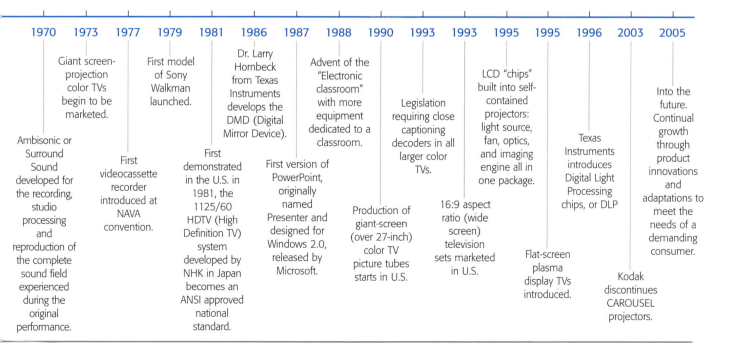

Audio-Visual Association. NAVA had one primary goal — to encourage increased government funding of education. This was accomplished with the passing of the National Defense Education Act in 1958. Billions of dollars went into America's public schools to improve educational opportunities for all citizens. This level of funding continued through the Kennedy and Johnson administrations, resulting in increasingly sophisticated methods of education for the public. The demand for more information drove the development of technological advancements.

Those first decades of rapid audiovisual development compare to the amazing pace of technological shifts in the last thirty years. The seventies saw the increased use of photography by the average person, the power of movie special effects, and the advent of computers in the business world. While not yet in every office, the computer age was not too far behind.

The '80s wrestled with a change of focus. While still used extensively for education, AV technology expanded beyond that arena to multiple markets. It marked the introduction of computers, digital audio and video processing and effects, new (and often confusing) videotape formats, satellite distribution, satellite newsgathering, fiber optics and high definition television. The byproducts of this feverish development were products that cost less but maintained performance. For example, improvements in computer microprocessor technology saw dramatic results in the features and performance of new systems. Consequently, equipment costs dropped and demand increased.

Also during the 1980s, there was increased interest in environmental concerns — both in the area of resource protection and the impact the environment can have on learning and communication. Acoustics, lighting and other building infrastructure have always been important in spaces where audio

and visual communications take place. With the advent of new technologies, the room environment and the physical infrastructure became even more important to AV design and integration.

In recognition of the global scope of these emerging technologies and the industry behind them, NAVA changed its name to the International Communications Industry Association (now InfoComm International). The AV industry had become a world player.

With the nineties came new display technologies, advanced microcomputers and the Internet, whose new media and business models have produced the most profound changes since Edison's moving pictures. The presentation market has boomed. LCD and DLP electronic projectors have become commonplace in corporate environments, higher education facilities and other presentation venues. The installation market has become more systems-oriented with sophisticated controls, network connections, videoconferencing and interactive technologies. Audio has become increasingly important, driven by consumers' demand for CD-quality sound in all application environments from classrooms to boardrooms.

Today, the industry remains committed to innovative technological enhancements, professional support services, and the appropriate integration of systems to ensure that end-users have the quality they need and want. Looking back, the audiovisual industry has developed and delivered the basics. Today, improvements in the quality of the presentation are paramount. "In many ways, AV technology has become more sophisticated, but it hasn't changed dramatically in the last few years," reflects Randal A. Lemke, Ph.D. who has served as Executive Director of InfoComm International since 2001. "Rather, we are focusing on the applications and the solutions to meet our end-users' needs. These solutions are what will provide the power to

Defining the AV Industry

Stated simply, the AV industry is dedicated to producing efficiently functioning AV systems that enhance communications throughout an organization or venue. It involves the technologies, products and systems for:

- Visual display
- Audio reproduction
- Video and audio recording, production and post-production
- AV signal control, distribution and transmission
- Lighting
- Control systems
- Interactive display and audio presentation systems
- Collaborative conferencing (video-, audio-, data- and web-conferencing)
- Streaming media
- Furniture and lecterns
- Cabling, connectors, racks, cases and accessories

This industry interfaces with the IT and telecommunications industries to create and transport the information that is presented to the intended audience.

The Essence of AV Design and Integration

Fundamentally, AV design and integration combines four basic but essential elements:
1. The *hardware*, or the physical electronics that are connected together.
2. The *software*, or programs that are loaded into the electronics to perform various AV tasks.
3. The *environment*, or the space and infrastructure within which the hardware and software work.
4. The *content*, which includes video, audio and data material created and presented using the software and hardware available.

Each one of these elements has specific and unique requirements that must be addressed during project design and implementation, and their ultimate interaction with one another will produce the desired solutions. Critical to successful project integration is the careful monitoring and balancing of the systems hardware, the software and the environments (or infrastructure), with the nature of presentation of the content in mind. Proper integration will ensure that the human aspects of AV (being able to clearly see and hear the presentation) are fully and successfully solved.

Segments within the Industry

The AV industry may be viewed simply on the basis of the outlets for the actual equipment. More specifically, differentiations are based less on products and more on the "relationship" that evolves from the services offered, the people involved, the procurement method used, and the desired applications.

---

[7] One could argue that Rental and Staging, where equipment is rented for short-term use, could be included in these divisions. This process includes installation and event management of temporary setups for special or ongoing events. For the purposes of this discussion of pro-AV system integration, it is touched upon but not fully explored here.

The industry serves and sustains relationships with three primary markets: consumer/retail, residential, and professional AV.[7]

- **Consumer/Retail**
  For the most part, consumer AV is comprised of personal electronics applications, such as DVD players, cameras and camcorders, boom boxes, home stereos, large screen TVs, and home audio and video devices.

  Consumer AV is almost always a direct relationship between the consumer and the manufacturer with a direct retailer (store or online) as the most common procurement method.

  In this industry segment, audiovisual devices are purchased as components or pre-packaged systems by the end-user from a retail establishment (or online outlet) and personally installed in his/her home or office.

- **Residential**
  For the most part, the residential market falls into the home theater or whole-house automation category. It usually involves a professional or company serving the residential electronics market and introduces the concept of systems versus boxes. Residential sales[8] represent a consumer relationship that most frequently goes beyond the retail establishment to a residential AV dealer that provides and installs the equipment.

  Home theater is targeted almost exclusively to higher-end residential installations, although there may be some overlap in small commercial applications. On small-scale projects, the procurement is the same as Consumer AV. On larger scale residential AV projects, the procurement comes as a "complete system" through a home theater/home automation specialist.

- **Professional AV**
  Pro-AV involves numerous specialists, but is almost always systems-focused. While box sales (a term used to define pro-AV equipment sold without dealer installation) are a large part of this industry, boxes are generally part of a larger, more sophisticated system.

  This industry segment includes commercial-grade equipment with some consumer or "prosumer" gear (equipment intended for use in both the general consumer market and professional situations).

  Pro-AV applications are in commercial, government, education, religious and transportation facilities, and include installations in arenas, stadiums, performing arts spaces, classrooms, auditoria, boardrooms, command-and-control centers, courtrooms, museums, training rooms, multimedia presentation rooms, call centers and many other venues.

  Other than high-volume end-user organizations that have direct sales arrangements, procurement is generally through a dealer/integrator. Because pro-AV involves systems, it often requires the design skills of an AV design professional and the installation skills of a

---

[8] According to InfoComm's 2005 Market Forecast Survey, both in the U.S. and other parts of the world, the increasingly sophisticated residential market is growing. Nearly 25% of U.S. dealers and 30% of European dealers are involved in residential sales.

systems integrator to advance beyond the box connection to the bigger picture of the system, its requirements and the overall installation.

What is Pro-AV Integration?

By definition, pro-AV integration is the process whereby professional audiovisual systems and equipment are designed, purchased and installed within a building project. Typical tasks include project needs analysis work, system design and engineering, equipment specification, equipment acquisition, cabling, equipment installation, interface and controls programming, testing, training and technical support.

Because overall project success is dependent on the physical environment in which the system is installed, the greater integration process includes coordination with building architecture, interior design, acoustics, lighting, structural, mechanical and electrical systems, both during design and construction phases.

There are other "low voltage" industries that are peripherally related to pro-AV, with systems that may involve audiovisual elements. These systems include, among others, security, nurse call, life safety, fire alarm, sound masking, computer and telecommunications (commonly referred to as data/telecom). Sometimes firms that provide pro-AV services also design and install these systems. However, there is a difference between pro-AV and other low voltage industries; each is a discipline unto itself.

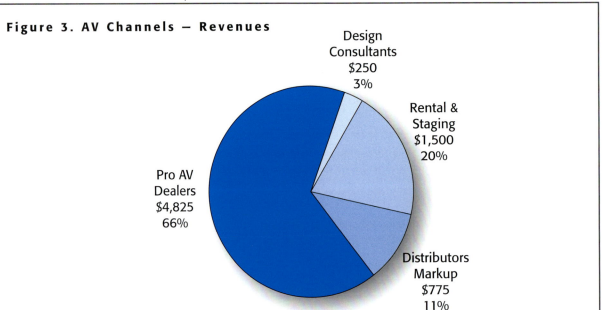

Figure 3. AV Channels — Revenues

- Design Consultants $250 3%
- Rental & Staging $1,500 20%
- Distributors Markup $775 11%
- Pro AV Dealers $4,825 66%

Pro-AV dealers, which include systems integrators and installers in this graph, are esponsible for close to 70% of the service component of the AV market size. Although the independent design consultants' aggregate revenue accounts for only 3% of the total service revenue, they drive a significant portion of dealers' systems integration revenue.

Source: InfoComm 2004 AV Market Definition and Strategy Study

The Demand for AV Professionals

As a direct result of a continually expanding marketplace, there is a corresponding need for AV professionals to help specify, integrate, install and service AV products for optimum performance. In addition, the importance of pro-AV is directly related to the growth and development of audio, video, control and communications technologies including personal computers and the Internet.

The continuing emergence of new markets and technologies brings new challenges to the systems designer in creating environments where images and sound are well-perceived by the audience.

In particular, the industry is experiencing growing pains as information technology (IT) professionals seek to gain knowledge about AV solutions and properly integrate them into the IT environment. New technologies such as streaming media, webcasting, digital signal processing (DSP), digital signage, wireless, and ever-developing software are among the areas creating shifts in processes and roles.

For those in the industry, pro-AV requires continually learning new ways of delivering value to customers. New technologies emerge daily; re-education and re-orientation is an ongoing state of affairs. To survive, AV professionals have to stay ahead of the game.

Those outside the industry often have had to learn the hard way. They find themselves in the position of having to make an AV project work without qualified professional guidance. At best, they will often spend more time and money than is necessary. At worst, they may have a system that does not perform to expectations.

The Audiovisual Solutions Provider (AVSP)

In almost every conceivable consumer or professional arena, there has never been a greater need for professionally designed, installed and produced AV presentation and communications systems.

An AVSP offers the professional guidance and technical expertise that most technical staffs within organizations do not have — or do not have enough of. If a company or facility is large enough, it may have an IT staff person (or department) in-house who will need the assistance of a trained AV professional to enhance the company's capabilities for the required system. Larger companies, government agencies, military facilities and educational institutions are more likely to have AV departments, with full time managers and technicians on staff. However, due to the typically large-scale needs of these major institutional clients, a qualified AVSP can be a great resource.

For smaller organizations without IT or AV staff, the need for outside AV professional services also exists. An AVSP provides the roadmap essential to making appropriate decisions about equipment rental and purchase, event production and the proper techniques for system installation and usage.

Regardless of on-staff expertise, an AVSP provides the most current information to enhance any organization's communications strategy. By virtue of training and experience, AV professionals understand and work with the complexities of IT systems and networked environments. Their involvement early in a project helps to guarantee that the in-house technical staff will know how to proceed without having to expend time and energy weighing options that may prove ineffective.

AV Resources — Who They Are

The following resources are available to any company or organization that is planning to add or upgrade its AV environments:

1. AV Manufacturers

AV manufacturers produce equipment used in pro-AV systems. They typically sell their products through channel distribution partners including dealers, manufacturers' reps and independent representatives. These dealers and distribution partners are trained by the manufacturers to market, demonstrate and offer products to end-users. Increasingly, manufacturers have also begun to sell products directly to end-users, creating market shifts and spurring changes in the value proposition presented by dealers and integrators to end-users.

2. AV Dealers, Distributors and Reps

AV dealers come in many different types and sizes. For the most part, they are local or regional companies or branches that are authorized resellers for the many different types of required AV equipment, although in some cases they may have a national or international focus. What distinguishes AV dealers from systems integrators is that dealers do not design or install systems, and are focused on selling the equipment. Some dealers offer equipment rentals and service over and above equipment sales. In that capacity, these dealers are considered rental and staging companies (see #6).

Because of their product training, AV dealers, distributors, manufacturers' representatives and independent representatives offer specific knowledge and "added value" when equipment is purchased. They communicate important information about the functionality of the equipment and its proposed integration in the facility.

3. Systems Integrators

AV technologies are often installed as integrated systems and require skilled installation professionals to ensure compatibility with existing IT systems and to provide staff training to work the system. The traditional role of the integrator has been to provide installation and wiring of physical systems. Increasingly the role of integrator has broadened to include a wider range of services, due to the complexity of the systems required. Some integrators now offer consulting and design services. When design, consulting and installation are combined and offered by a systems integrator, the service is referred to as the "Design/Build" process.

4. Independent Design Consultants

Often hired by architects or directly by system owners, the independent design consultant determines the best options for the design of a system and produces specifications for the job that can then be competitively bid by qualified systems integrators or negotiated with a preferred integrator. Handling video, audio, data and PC-video, as well as designing the acoustics, lighting, electrical power and other infrastructure appropriately in a room requires the expertise of a system designer who understands complete system functionality.

Design consultants act as the owner's technical representative not only to ensure that systems integrators are performing the job according to professional specifications, but also to handle issues that may arise during construction. Because they are not affiliated with any particular brand or manufacturer, design consultants provide independent advice on system components.

There are also business consultants within the industry who do not offer design services; their services may include strategic development, marketing, project management or multi-media creative services.

5. Independent Programmers

A growing segment within the AV industry, independent programmers are individuals or companies that are not employed by hardware vendors or integrators, but rather provide them programming expertise and service for hire. Independent programmers are not only involved in control system programming, but also in any specialized software to provide a working system. Many, through their work with consultants and integrators, also have design and installation experience.

6. Rental and Staging Companies

These companies offer event staging and AV technology rentals, including production services for corporate meetings, conferences, conventions, trade shows, product launches, special events, entertainment and concerts. For these events, AV professionals are on hand for the duration of the event to set up, take down and operate projectors, cameras, screens, sound systems, scenic elements, lighting and special effects.

7. Presentation Professionals

After hardware is installed in an AV system and staff training is complete, presentation professionals are often hired to produce strong, effective and creative presentations that use the AV equipment to its fullest potential. This ensures a solid return on the organization's investment. Presentation professionals are often hired as in-house resources or as independent freelancers to enhance the message through graphics and video presentations, marketing expertise and creative application of presentation tools.

8. Production Studios

This category covers a broad range of facilities and services, such as video production, staged events, scripting, special effects and coordination with rental and staging companies.

What is AV Industry Certification?

Studies indicate that many individuals who enter the industry do so from other fields. Often, individuals with careers in music, art, theater, engineering or computers become interested and involved in the AV industry. A professional education and certification program allows these individuals to learn the technical aspects of the business and to demonstrate their AV technical proficiency to colleagues and clients.

With the unique requirements of designing, installing and operating AV systems, AV pros require special training. InfoComm International offers a wide variety of education options, both online and in classroom format, from the most basic to advanced levels of curriculum. Training covers numerous disciplines and topics, including AV design, installation, networking, audio, video, conferencing and other subjects.

In 1981, InfoComm recognized the need for a strong certification program that would validate the credentials of its professionals, and established the Certified Technology Specialist (CTS) program.

The first step for individuals is to take the General CTS test, which is offered online. Individuals can advance to specialized testing in Design (to achieve the CTS-D) and Installation (to achieve the CTS-I).

InfoComm International is also the only association to offer a company-level certification, the Certified AudioVisual Solutions Provider (CAVSP) designation.

The CTS credential is for audiovisual professionals who have demonstrated knowledge and comprehension of the science and technology used in communications including audio, video, display and systems. By demonstrating their knowledge of how things work, a CTS is able to apply these fundamentals to new technologies and applications. To achieve a CTS, the individual must also agree to a code of ethics pledging truth, accuracy, and a commitment to excellence in all aspects of the profession. InfoComm's general certification is recognized by the National Certification Commission.

The InfoComm CTS-D credential is the specialized certification for audiovisual systems design professionals. A CTS-D is an individual who earned this specialized certification by demonstrating detailed knowledge of how to analyze, select and plan seamless audiovisual communications equipment interoperation. This knowledge is demonstrated by the CTS-D in the form of complete specifications and drawings that deliver a desired outcome to meet a client's needs. A minimum of two years of industry experience is required in addition to the theoretical and practical competencies. To achieve a CTS-D, the individual must also agree to a code of ethics pledging truth, accuracy, and a commitment to excellence in all aspects of his or her profession.

The InfoComm CTS-I credential is the specialized certification for audiovisual installation professionals.
A CTS-I is an individual who earned this specialized certification by demonstrating skills and knowledge through a rigorous regimen of testing in a broad array of installation practices and techniques. A CTS-I is proficient in installing all components of a professional audiovisual system. A CTS-I takes on additional responsibility for projects to ensure quality, efficiency and safety on the job site. To achieve a CTS-I, the individual must also agree to a code of ethics pledging truth, accuracy, and a commitment to excellence in all aspects of his or her profession.

The InfoComm CAVSP credential is a company recognition awarded to organizations who have demonstrated their commitment to professional excellence by supporting their employees who achieve and maintain individual Certified Technology Specialist CTS credentials. Gold certification recognizes 75 percent of all sales, customer service and technical staff have and maintain InfoComm certification, while silver certification recognizes that 50 percent of staff are InfoComm certified and basic certification recognizes that 25 percent of staff are InfoComm certified. CAVSP companies have also agreed to comply with 10 Standards of Excellence that were developed in collaboration with industry experts to emphasize the best practices of AV businesses. The Standards of Excellence are a guidepost for AV companies and professionals who place the customer first and offer quality AV solutions. They include the commitment to provide complete customer satisfaction with AV solutions that provide value to the client. These standards convey the importance that AV solutions providers place on their own skills development through continuing education. They also cover the critical role that AV professionals play as partners with IT specialists, architects, building managers and other building trades.

Nationally Recognized Certification

In 2005, InfoComm became the only AV association to receive national registration for its certification program. Awarded by the National Certification Commission, this recognition was given after a review process considering 20 criteria and guidelines for effective administration and presentation of certification programs. Examples of these criteria include adequacy of the program to recognize the career in an assessment of education and experience; process availability and objectivity; appropriateness of the designation title; number of approved participants; staff resources and functions; advisory committee composition, qualifications, and responsibilities; and applicability of components, requirements, and the evaluation.

To help end-user customers avoid unqualified providers, InfoComm established an international awareness campaign designed to promote the AV industry and to direct clients to qualified AV firms. Hundreds of providers have committed to the program and many have become Certified Audiovisual Solutions Providers (CAVSPs).

A Word on Other Credentials

It should be noted that there are a few other credentials that may be encountered in pro-AV and related fields. The "Professional Engineer" (PE) license is primarily used for mechanical, civil, electrical, structural, plumbing and other essential building industry trades with life-safety concerns. AV firms sometimes employ PEs in mechanical, electrical and acoustical engineering, which may not be AV-specific. PE licensure allows firms to use the term "engineer" in their company names and to be listed in various engineering resources. The Registered Communications Distribution Designer (RCDD) certification, which is issued by Building Industry Consulting Service International (BICSI), is targeted toward data network and communications professionals, although this is becoming an essential aspect of AV work. The ISO 9001:2000

Standards of Excellence

AV Solutions Provider (AVSP)

InfoComm administers an awareness campaign designed to help customers understand the AV industry and select qualified AVSPs to perform the work. AVSPs commit to the following 10 Standards of Excellence:

1. **Complete Solutions** The AV Solutions Provider (AVSP) works to provide a comprehensive AV system that meets the client's communications needs.

2. **Informed Advice** The AVSP pursues a collaborative relationship with the client to understand the needs and recommend solutions to achieve complete customer satisfaction.

3. **Best Value** The AVSP seeks to give the client the best possible value and return on investment.

4. **On-Going Technical Support** The AVSP provides on-going support for the AV system through warranty and maintenance programs to ensure the system's usability and the client's satisfaction.

5. **Systems Compatibility** The AVSP designs or provides a seamless operational interface for components used in a rental environment or permanent installation to create a stable and viable system.

6. **Scalability** The AVSP designs and integrates AV systems that promote ease-of-use, long-term cost-effectiveness and upgradeability.

7. **Clear Scope of Work** The AVSP specifies in the proposal and quotation all the necessary components and services required to create the AV system and clearly calls out and discusses any exceptions.

8. **Appropriate Documentation** The AVSP provides appropriate and complete documentation of the system as specified in the vendor's proposal and quotation or the consultant's specification.

9. **Expert Technical Staff** AVSP staff are in programs to become industry certified, to maintain their certification and to keep current on new developments in AV technology.

10. **Training Support** The AVSP is a source of professional training services to help the client become proficient in using AV technologies and systems and to advance their communications goals.

Audiovisual Best Practices

AV Client Commitment

Recognizing that on the client side, success might be achieved through a similar level of commitment, the Technology Manager/End-Users Council of InfoComm determined that there should be a parallel set of standards to which end-users could aspire. A strong commitment to these standards reveals a corresponding success rate in the final outcome of AV system use, and they are offered here as a way to establish the dialog with end-users throughout the process of design and integration.

1. **Complete Solutions** AV clients know their communications needs or can assemble the parties necessary for Audiovisual Solutions Providers (AVSPs) to provide a comprehensive system to meet these needs.

2. **Informed Advice** AV clients pursue a collaborative relationship with AVSPs to consider cost, performance, schedule and expectations for new projects, ongoing operations and maintenance programs.

3. **Best Value** AV clients have business plans that address communications needs, opportunities and resources to achieve the best value from their AV products and services.

4. **Ongoing Technical Support** AV clients actively engage with AVSPs to determine and apply the appropriate resources to maintain the client's AV systems, considering a range of onsite or remote support and training available through AVSPs and manufacturers.

5. **Systems Compatibility** AV clients alert AVSPs to existing standards and practices that will affect AVSPs' delivery of seamless and consistent operational interfaces.

6. **Scalability** AV clients routinely evaluate their current and future communications needs and resources to assist the AVSP in providing AV systems that are useful, cost-effective and conducive to upgrades.

7. **Clear Scope of Work** AV clients provide AVSPs with the specifications that will affect the scope of work and carefully evaluate the AVSP proposal or quotation to confirm that the project is effectively addressed.

8. **Appropriate Documentation** AV clients clearly define document requirements that are proportional to the sophistication of given AV systems, the capabilities and needs of end-users and support staff, and the service expectations that the system must meet.

9. **Expert Technical Staff** AV clients are committed to the ongoing training and certification of technical and operations staff to keep current on new developments in AV technology.

10. **Training Support** AV clients actively seek the appropriate level of training support from their AVSPs based on the clients' business needs, in-house expertise, location and other factors.

certification is based on standards developed by the International Organization for Standardization (ISO). While not AV-specific, it is a recognized certification that acknowledges attention to management and quality control performance within a company.

The Need for Collaboration

The mission for the AV professional is to create audiovisual environments that work — and work well. This is more difficult today than ever before.

AV is no longer an add-on to the building project. AV communications systems are increasingly a part of today's building types. They are critical to the workflow and success of the environments we hope to create. AV technology use varies from simple room signage to full mission-critical operation communications centers.

In the days of overheads and slides, AV professionals dealt almost exclusively with other AV pros or end-users (who were also technicians). In a pro-AV integration project today, people from different disciplines are involved — many who are unfamiliar with AV "techies." In addition to AV consultants, integrators, manufacturers and technicians, an AV project also combines the skills of architects, engineers, general contractors, subcontractors, building facility managers and a variety of specialty consultants.

Building design and construction has become more complex and expensive. The course of a pro-AV project (as a low voltage trade) now often has an impact on construction projects equal to or even larger than any other single trade. That means the AV professional has (or should have) direct involvement in all aspects of building design and construction, including space planning, sightlines, lighting, acoustics, electrical and mechanical systems. Quite simply, it means that the AV professional needs to relate to almost everyone who works on a construction project that involves an AV system.

The work of AV professionals during the course of an AV integration project has more impact than ever before on the other building industry designers and contractors. This dynamic requires teamwork. The creation of a technology-enhanced building requires a team made up of professionals from the AV industry and the other disciplines. A critical part of that team effort is learning about the different professions and the language they speak.

In many ways, contemporary AV integration is really more about project management and procedure than it is about audio and video or design and construction. It is about ways in which a project is handled and how each of the various professionals performs a critical role in the implementation of an AV system. Most, in fact, are there from start to finish and, despite their various job functions, are members of the same team.

The AV industry depends upon this teamwork, or *collaborative project process*, that results not only in satisfied clients (the owners and end-users), but also fosters the level of professional involvement that improves the industry as a whole.

OVERVIEW
of the Design and Construction Process

OVERVIEW

Because pro-AV design and integration occurs within the building design and construction process, these chapters are organized to correspond to this architectural sequence. Highlighted here are important elements such as how pro-AV fits into the process as well as best practices for the AV professional, the architectural and engineering design team, the building construction team and the facility owner. It is a start-to-finish approach with recommended practices and suggestions for handling issues that arise during the project.

The chart below provides an overview of the two primary AV system delivery methods: AV consultant-led design-bid-build and AV integrator design-build. The processes are shown left to right as a rough timeline in relation to the overall building design and

—Continued on next page

There are times when the phases listed here are compressed within a small project; however, large projects generally follow these steps if they involve an architectural firm in the process. These phases in sequence are:

Program Phase
Within the building design and construction process, the term "program" is used to refer to the needs analysis phase. In this phase, the users' project goals and needs are translated into a textual report of building design criteria, occupancy, use, space program (types and sizes), space adjacencies, project schedule and project budget.

Figure 4. AV Process Overview

Project Process	Contract Building Design Team	Building Program and Budgeting	Schematic Design	Design Development	Construction Documents	Bid & Contract Base Building Construction Team
Consultant-Led Design-Bid-Build	Select AV Consultant RFQ/Shortlist/RFP Contract for Design (see Chapter 3)	Develop AV Program and Budget	Preliminary AV Systems Design	Design Base Building Infrastructure for AV		Review Coordinate AV-Related Base Building Design Elements
Integrator-Led Design-Build	Select AV Integrator RFQ/Shortlist/RFP Contract for Design (see Chapter 3)	Develop AV Program and Budget	Preliminary AV Systems Design	Design Base Building Infrastructure for AV		Review Coordinate AV-Related Base Building Design Elements
Owner Duties	Provide Input to AV and Building Program			Review Base Building Design		

Chapter 4: The Program Phase

Chapter 5: The Design Phase

Design Phase

The design phase translates program information into drawings and specifications. It is comprised of the following sub-phases:

- **Conceptual Design Phase:** Following architectural programming, the architect sometimes creates a conceptual design – a one-line diagram that graphically portrays the program information for space shapes, adjacencies and sizes.
- **Schematic Design (SD) Phase:** The conceptual design is developed to a more detailed level, beginning to show more detail such as double lines for walls, door locations and room orientations. In addition, the architect defines the overall "massing" of the building(s), and a schematic narrative generally describes the major systems to be included in the building.
- **Design Development (DD) Phase:** The goal of design development is to move beyond major coordination issues to the basic floor plans. During this phase, all major design decisions are made and finalized with the owner so the building floorplan is set, engineering systems selected and detailing can commence. This is an intense period of design consulting and decision making for the design team and the owner. The end result is the final architectural and engineering design.

—Continued from previous page

construction process with key tasks shown for each phase. Reflecting the owner's crucial role, certain key owner tasks are also shown. The many tasks performed in the AV design and integration process are discussed in detail within each chapter.

Reading vertically, an overview of the construction and related AV tasks can be found by project phase, with the chapter associated with each phase noted at the bottom. Some aspects of the process that may occur in one phase but are described in another chapter are noted in the contrasting boxes.

Base Building Construction			Building Commissioning	Building Occupancy	
	AV Bid		AV System Installation		
Develop AV System Design Package (see Chapter 5)	Select AV Integrator RFQ/Shortlist RFP	Review AV Bids	Contract Integrator (see Chapter 3)	Pre-Test System	Commission the AV System	Train the End-Users	Warranty Period Begins
Review AV-Related Base Building Submittals	Monitor AV-Related Base Building Infrastructure Construction		Review AV Systems Submittals				
Contract Same AV Integrator for AV Installation (see Chapter 3)	Develop AV System Design		Pre-Test System	Commission the AV System	Train the End-Users	Warranty Period Begins
Review AV-Related Base Building Submittals	Monitor AV-Related Base Building Infrastructure Construction					
Review AV Bids	Provide OFE	Provide Communications Services		Coordinate End-Users and Spaces for Training	Sign-Off
Review AV Systems Design including Graphical User Interfaces		Review AV Systems Submittals			

Chapter 6: The Construction Phase

Chapter 7: Commissioning and Training

Overview

Within the architectural design process, the design development phase is a go/no-go decision point. Usually, enough design information has been gathered by the end of this phase to know with a fair degree of accuracy how much the facility construction will cost. For some projects, this is the stage at which the owner may decide that the project should be abandoned because of budget or other issues before proceeding into the construction documents phase.

Construction Documents

The goal of this phase is to create the documents from which the facility will be built. Projects may range from a small room renovation to a skyscraper or multi-building development.

This is a coordinating and documenting phase rather than a design phase. The goal is to define the design in a manner in which it can be purchased; the result is drawings, specifications and contracts for the building's materials and construction.

Bidding

The bid phase occurs when a design-bid-build process is used. It is the period when the construction documents are distributed with additional contract language to potential bidders. This package is often referred to as the "contract documents," and becomes the bulk of the actual contract between the building owner and the contractors hired to renovate or construct the building.

Construction Phase

Once the contract is signed, the construction phase begins, during which the actual site work is conducted, the building is constructed or renovated, and the engineering systems are installed.

Commissioning Phase

When the construction work is at or near completion, there is often a commissioning phase, during which the building systems are put into operation, tested and optimized for their operation in the building.

Warranty Period

Once the building is complete and commissioned, the warranty period begins on the building and its engineering systems. This is typically a one-year period during which, just as with the warranty on a car or major appliance, the contractor is required to correct any problems that are a result of the "manufacture" of the building.

After introducing the people, organizations and roles that participate in this process, we will again review the design and construction phases, this time from an AV point of view. As the process is addressed in more detail, important questions arise that affect the complexity and the quality of the entire AV design and integration process. They are:

- What are the AV-related concerns for the design and installation teams?
- When should the AV provider get involved?
- At what points during the process should the AV provider be included?
- Who is the AV designer working for?
- Who is the AV integrator working for?
- How do the AV systems and infrastructure fit into the overall project budget?

These questions are addressed throughout the book, let's begin with a look at the project teams and how AV affects each group. In each instance, there is overlap in areas of responsibility; communication is key to success. The description of the project team participants also brings up specific issues that crop up frequently during the course of a building project where AV is a component.

CHAPTER 1
Understanding the Project Team

CHAPTER ONE

Professional audiovisual system design and integration are now more crucial to the building industry than ever before. Before the development of modern pro-AV integration, most AV professionals dealt almost exclusively with other AV people or with the end-users of AV products such as technicians or presenters.

In a pro-AV integration project today, AV and building industry professionals work as a team due to the complexities and the "built-in" aspect of the contemporary AV project. Professional encounters among the architectural, engineering and AV trades have created a need to disseminate information to all involved on who is doing what.

Chapter 1 covers the four teams representing the range of common roles within the design and construction process: Owner, Design, Installation and Management. The role of AV pros – as well as their relationship to the more traditional building trades – is highlighted.

The chapter works to create a better awareness of project roles and responsibilities and to indicate how good communication is the key to a successful project.

An Overview of the Teams

Before reviewing the process and procedures, it is important to look at the roles of participants within the process. There is a host of individuals that one is likely to encounter, and each has a role to play in the path to architectural, as well as audiovisual, success. In addition, there are multiple paths to success that will be explored in later chapters.

The project participants are usually thought of in terms of teams which are comprised of a number of groups and individuals. Depending on the scope of the project, these professionals and their firms can be classified as follows:

Figure 5. Project Team Overview

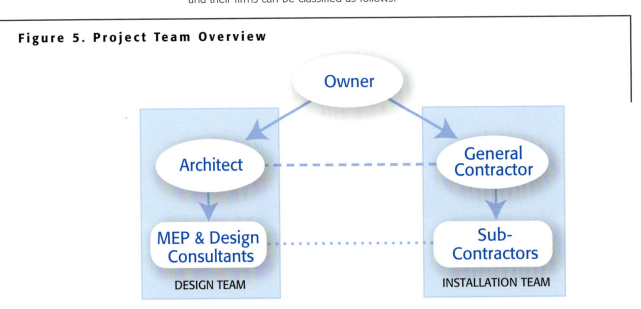

LEGEND
→ Contract Relationship
--- Coordination commonly required by contract
···· Additional coordination commonly practiced

While coordination is required among members of the design team, contracts may require that all communication between parties flow through the contracted or sanctioned lines of communication indicated here by the dashed or solid lines.

The Owner Team

This is the entity or entities involved as the buyer of the AV systems. The owner team may include several groups or individuals who participate in the project process:

- End-user
- Facility Manager
- *AV Technology Manager*
- Building Committee
- Buyer, Purchasing Agent or Contract Representative

The Design Team

This group designs the building and the systems and may include the following groups or individuals:

- Architect
- *AV Designer**
- Interior Designer
- Mechanical Consultant
- Electrical Consultant
- Plumbing Consultant
- Structural Consultant
- Lighting Consultant
- Data/Telecom Consultant
- Acoustical Consultant
- Security Consultant
- Life Safety Consultant
- Other Industry and Trade-specific Consultants

The Installation Team

This group provides construction and installation services and may include the following groups or individuals:

- General Contractor
- *AV Integrator*
- Mechanical Contractor
- Electrical Contractor
- Plumbing Contractor
- Structural Contractor
- Lighting Contractor
- Data/Telecom Contractor
- Security Contractor
- Life Safety Contractor
- Other Trade-specific Contractors

The Management Team

This group provides management services on the project and is usually associated with or represents the owner in some way.

- Developer
- Construction Manager
- Building Management Agency
- Move Consultant

Each team is reviewed in detail along with its role in the process and how each relates to AV.

---

* This can be either an independent AV consultant in a consultant-led design process or a designer working for an AV integrator in a design-build scenario.

The Owner Team

Depending on the type and size of the owner organization, participants may consist of a small number of people or a large group. An Owner can be a corporation, a government agency or an individual. In some cases, the owner of a building is fitting out a space to be leased to another company that is the actual purchaser of the AV systems. Some owner configurations are more complicated, such as in some state university systems, and some are simple, as in the case of a small private company such as a law firm or a small manufacturer.

The owner typically has its own representatives in the building design process. Representatives range from non-technical administrative employees to CEOs and from facilities managers to in-house architects if the client owns or leases a lot of property. Sometimes a team or committee represents the owner.

Listed below are components of the owner's representation in a typical pro-AV project, along with types of concerns related to the impact of AV on their role.

End-User

End-users are the people who will ultimately use the systems and facilities that are being designed and installed. It is these individuals and groups for whom the project is being built and who may be responsible for paying the bill for the provider's services.

It is important to note that the end-users may not be employed by the owner's organization, although in most cases they are. In addition, later mentions of the owner may or may not represent the end-users, depending on the structure of the project.

In terms of AV systems, there are generally two types of end-users — the technical end-users and the non-technical end-users:

Technical end-user

In the AV industry, technical end-users, also known as technology managers, are responsible for operating, setting up, maintaining, scheduling and managing the AV system from a technical standpoint. Most are familiar with the IT network and how the AV systems relate to it; some are also responsible for the IT systems.

Non-technical end-user

The non-technical users are sometimes referred to as the "real" end-users. These are the presenters and meeting participants who use AV systems for their intended purpose: information communication. These may be teachers, trainers, students, salesmen, CEOs or anyone who participates in an event that utilizes an AV system.

Facility Manager

Larger organizations have a staff facility manager heading up a department that covers everything from janitorial services to renovation work and from planning to operations.
Facility managers are concerned with physical facility standards (standard finishes, cabling schemes and electrical/mechanical services) and construction schedules from the owner's operational perspective. Their areas of AV concern may include space allocation, cabling standards and installation schedules, since large AV installations often overlap with occupancy periods once the general contractor has finished work.

BEST PRACTICE

Keep the AV manager involved

The staff AV manager is very involved with the details of the AV design and installation. The AV department establishes the owner's AV standards for the project. The manager is also knowledgeable about the needs of the end-users in relation to the current systems. The AV manager is motivated to ensure that the project goes smoothly, with communication being a vital part of that success. Consequently, good communication between the AV manager and the AV providers is critical.

AV Department Manager

Most organizations have someone on staff who manages the current AV systems, schedules usage of the facilities and maintains the facilities. Smaller organizations may not have a dedicated position, and the responsibility resides with someone who has the closest level of experience or expertise, e.g., someone in the IT department.

In some instances, there is an experienced AV manager who is a knowledgeable technical representative. This manager also coordinates training for the system upon completion of the installation — both for the end-users and for the technical support team. At times, the AV technology manager may act as the owner's representative or as the point of contact for coordination of the AV project.

Building Committee

Sometimes, a project involves several divisions or departments involving a number of administrative and technical stakeholders in the project. In this case, the owner may create an internal committee to steer the process of building the new facility. This type of committee should have representation from the technical staff who can provide valuable input to the design process.

Buyer, Purchasing Agent or Contract Representative

Medium to large organizations often have a department or individual whose dedicated role is to manage contracts, select vendors and establish or negotiate contract pricing and terms. These personnel may be involved in the construction document preparation (particularly for the administrative portions), in the bidding and contract negotiation process for both designers and installers, and in administration of billing, paying and contract close-out tasks.

The Design Team

Once the owner has made the commitment to a new building or a significant renovation, the designers begin their work. This team consists of the architect, interior designer, engineering and other consultants and, of course, the AV designer.

In addition to the architect, other designers are critical players in the construction of the building. The most typically required roles are mechanical, electrical and plumbing (these three together often referred to as "MEP") and structural, assuming there are proposed structural modifications. These professionals may be associated with a multi-disciplinary firm or part of a design-build entity (see next section for detailed discussion). Other specialized consultants may also be a part of the team.

Architect

The architect is responsible for the design and functionality of the building to meet the owner's programmatic needs, aesthetic expectation, and budgetary requirements. The architect is also responsible for the life safety and building code compliance of the project. Generally, the architect's firm holds the "prime" contract for the building and the technical system designs, subcontracting other design consultants under its contract.

The architect's role can be filled by individuals from a wide variety of practices including individual architects in practice (standard architectural firms), combined with engineers (either A/E or E/A firms), combined with contractors (design-build firms) or as licensed representatives of the owner's own staff. There may be two architectural/engineering design team entities if the complexity of the project warrants it. This is common for large projects where there may be a design team for the base building and another for the interior fit. Often one team represents the building owner and the other the end-user.

BEST PRACTICE

Create a technical subcommittee

To ensure that a building committee has a technical voice, an enlightened owner may assign a separate subcommittee to allow the various technical representatives to work together and represent the owner as a group. This allows the AV providers to work through a point of technical responsibility for the owner and provides a venue for the owner to resolve conflicts, duplications or other issues internally without getting the AV providers involved in internal politics.

The committee is essentially concerned with the same issues as an individual AV Manager, perhaps with broader interests if data/telecom, security or other interests are represented. The committee has information about standards (which may first require definition), as well as about the end-users' needs. In addition, the committee can convey concerns about AV systems interconnections, operations and maintenance that will have an impact on the systems design.

BEST PRACTICE

Involve technical consultants early

Since AV costs can be a crucial component in the contract, it is recommended that the architect confer early in the process with the AV designers to determine what impact the proposed AV systems will have upon the building plan and budget. AV systems considerations include acoustics, lighting, sightlines, and space requirements that may require design adaptations. Examples are additional space to accommodate projection rooms, thicker wall construction to enhance acoustics, additional energy requirements and budget for lighting systems, and, in some cases, a taller building to allow for increased floor-to-floor dimensions required for optimal sightlines. It is critical that the AV designer lead the client to timely decision making on behalf of the overall design team, whose work depends on this information.

Interior Designer

The interior designer is closely associated with the architect and may even be the same individual performing both roles. The designer is most often concerned with the furnishing and appearance of the room: how the wall, floor, ceiling and furnishings are finished and situated. He or she advises on or decides if surfaces are to be painted, fabric-covered, tiled or textured and what product, material, colors and patterns are to be provided. Sometimes these finishes and the furnishings have a direct impact on the AV system. Accommodations may need to be made for wiring, microphones, switches, AV connectors, monitors, control panels, projectors and other AV devices.

Mechanical Consultant

The mechanical consultant designs the heating, ventilating and air conditioning (often referred to as "HVAC") systems. This consultant works with the architect to determine the mechanical system needs for the building and creates the designs that include the chillers, pumps, fans, piping, ducts, diffusers, and grilles that heat, cool and ventilate the building. Often, the mechanical consultant deals with building automation systems that include computer-based monitoring, control and diagnostics of the mechanical system. Since HVAC systems tend to be noisy, and AV technology systems require special attention to avoid over-heating, it is critical that the AV system design is taken into consideration when designing the HVAC system.

Electrical Consultant

The electrical consultant designs the "high voltage" power distribution systems in the building. This includes the basic electrical power systems from the power company's entrance to the facility, following through to the power outlets in the building. The electrical consultant often designs and/or coordinates the conduit and cable tray designs that support the "low voltage" systems, such as data/telecom, AV, life safety and security systems.

Specifically for AV considerations, high voltage systems may need to include an isolated ground system for AV-related outlets, transient voltage and surge suppression (TVSS) systems, uninterruptible power sources (UPS), and acoustical treatments for the building's power transformers.

Plumbing Consultant

The plumbing consultant is responsible for the piping associated with water and waste into and within the building. This includes sprinkler piping, chiller water for the HVAC system, and natural gas service. The acoustical design of the building may affect the plumbing design by requiring special mounting or rerouting of piping around acoustically sensitive areas.

Structural Consultant

The structural consultant is the designer for the building structure. This involves fundamental decisions, such as whether the building will be formed concrete or steel frame and the size, spacing, and placement of beams, joists or columns throughout the building. Though the structural engineer may not deal directly with the AV designer, some of the issues related to sightlines can affect the column spacing and even the floor-to-floor structural clearance that are fundamental to the building design. Likewise, the structural loads of large AV equipment, such as speaker clusters and ceiling-mounted, large-venue video projectors and lifts, need to be factored into the structural design.

The AV professional must also work closely with the structural consultant to collaborate on the design of any exterior signage installations attached to the building, roof top equipment installations (such as satellite dishes and antennas), structural supports required for equipment hung from the building structure (e.g., speaker clusters, ceiling-mounted, large-venue video projectors, screens, and lighting grids), and the creation of cable/conduit pathways through structural floors.

AV Designer

Throughout the project process, the AV designer is involved with analyzing the end-users' needs and translating them into infrastructure and systems designs. AV designers must be focused on coordinating and monitoring the process from start to finish to ensure the AV system's success. To that end, the AV designer is the source of collaboration and coordination for all other consultants on the design team.

During the design phase, the AV designer can be involved in two forms:

Independent AV Consultant
The independent AV consultant works in much the same way as a mechanical or electrical consultant, providing design consultation to the installation process without providing any of the equipment.

AV Designer within a Design-Build Firm
Under an AV design-build process, the AV designer may be part of an AV integration firm that provides the infrastructure design, the system design, and the system installation. Also known as AV systems integrators, these firms also may provide the AV equipment required to outfit the facility.

Less frequently, the AV designer is a member of the owner's organization. In either circumstance, the AV designer is equally responsible for the same range of design issues. From an infrastructure standpoint, these requirements include conduit, power, data, lighting, acoustical[9], space plan, sightlines and HVAC requirements. The AV designer also handles the design and cost estimating for the electronic systems.

During construction, the independent AV consultant monitors construction of both the building and the electronic systems. At the end of the project, the AV consultant or designer often assists with the owner's technician and end-user training.

---

[9] While many independent AV consultants can provide acoustical consulting as a part of their scope of services, this specialized area may also be provided by a firm dedicated to this field.

BEST PRACTICE

Involve the AV designer early to coordinate building structural design

Sufficient ceiling heights are critical to AV sightline concerns. For example, if a very large audience area is planned, the required image size and location may necessitate raising the ceiling, altering the floor structure, or even increasing the height of the entire building to allow for AV presentations. To avoid the budgetary and disruptive consequences of these changes, it is important to discuss these concerns with the AV designer early in the process.

BEST PRACTICE

Involve electrical consultants early

Discuss the AV operations with the building design engineers early. Coordinate with the mechanical engineer to determine if dedicated HVAC systems or services will be required for specific rooms and equipment. Confirm the hours of operation of building HVAC systems. AV equipment may require 24x7 HVAC services which would require separate HVAC controls or services. Coordinate with the electrical engineer to determine the reliability requirements for electrical power and communications servicing the AV systems. Mission-critical AV and information systems may require dedicated and redundant power systems. Remember that supporting HVAC systems, lighting and life safety systems must have redundant power as well to maintain the operational environmental conditions.

The relationship between AV and IT systems

Since AV communications systems have become integrally tied to IT systems[11], there is an even greater need for advising the data/telecom consultant what data, telephone, ISDN, and other communications outlets are needed for the AV systems beyond basic requirements for the end-user. In addition, the data/telecom consultants will include fiber and possibly cable TV cabling in their design that will be utilized by the AV contractor for transporting AV signals. These need to be coordinated between the AV designer and the data/telecom consultant during the design phase of the project. The AV system may utilize the structured cabling system either for local networking (e.g. control systems and audio or video distribution), Internet access (for system monitoring, videoconferencing or multimedia content access), or telephone and ISDN access (for audio and videoconferencing). Outlets may be installed in dedicated data/telecom outlets or in back boxes and plates specifically for AV use. These issues must be coordinated between the data/telecom and AV contractors before and during installation and system testing.

Lighting Consultant

The lighting consultant is responsible for determining the required lighting for each space and must provide layout and specifications for the lighting system for each area. This includes the fixture types, lamps, wiring and control features (switching and dimming) associated with each space.

Close coordination is required between the AV designer and the lighting designer. Whether it is a basic presentation space with direct-view monitors, a training space with front projection or even a videoconferencing system with cameras and multiple displays, lighting plays a critical role in the quality of the visual aspect of the audiovisual environment. It is crucial to the AV system's success. Another important consideration is that the AV system may require interfacing to the lighting control systems.

Data/Telecom Consultant

Data/Telecom includes the design of the cabling and physical infrastructure (sometimes called the "structured cabling system") that accommodates local area computer networking, Internet access, telephone systems, and other communications systems. Typically, the scope of work includes the cabling, space planning, and associated infrastructure requirements, such as conduit and power requirements[10].

Acoustical Consultant

The acoustical consultant designs the building components that affect sound isolation, reverberation time and noise reduction as they relate to every physical element of the building, including ceiling, wall and floor finishes, wall and floor/ceiling constructions, window selections, fan and chiller selections, duct and pipe routing, equipment mountings and room shapes. The acoustical consultant needs to know what spaces are needed to house the AV systems as well as the type of system. Working closely with the architect, interior designer, HVAC engineer, and the AV contractor, the acoustical consultant can properly establish the appropriate acoustical criteria for each space and provide recommendations.

Security Consultant

The security consultant deals with audio and video equipment such as cameras, microphones, recording equipment, and video displays as part of the security designs. These systems are very specialized and include barrier construction, door lock and card key systems, and owner procedural issues.

If the security system is not completely separate from the AV system, there may be crossover where cameras, microphones, video displays or intercom systems are positioned. The AV and security consultant need to coordinate their system designs and infrastructure to ensure that they are operating in alignment.

Life Safety Consultant

Life safety systems provide emergency alert to building occupants. The most common design issues are related to fire detection, occupant alerts or hazardous substances, such as natural gas and refrigerant leaks.

The life safety consultant needs to know where the AV systems are located in the building to ensure that the emergency life safety systems (i.e., lighting, public address, fire alarms, sprinklers and egress) are coordinated with the AV systems and that the appropriate lighting and audio control interfaces/overrides are in place. This is critical in all public assembly spaces and venues.

---

[10] Today, some pro-AV companies are also providing data/telecom consulting to their clients.
[11] In fact, today the data/telecom and AV consultant are more frequently employed by the same company.

Other Industry- or Trade-specific Consultants

On any building project, numerous other "industry-specific" consultants may be involved, especially when the building requires specialized systems, such as a restaurant, hospital, lab or theater. Even standard office buildings may involve consultants for door hardware, landscaping, or civil engineering. The AV design may require coordination with these consultants depending on the type and location of the systems that are needed.

The Installation Team

After the designers convert the owner's requirements to paper (and electronic documents), the installers/contractors begin their work. Installation contractors are often firms solely dedicated to installation. However, in a design-build process, the designers and installers may be part of the same company. In some cases, a "fast-track" process is used in which construction starts before some of the designs are finished (e.g. structural designs are completed and construction begins before the interior space plans are designed).

For all members of the installation team participating in the construction phase, coordination is key. The pro-AV providers need to coordinate closely with the traditional base building trades to make sure that the infrastructure for AV is being installed according to the design and that any field conflicts or questions that come up are correctly resolved. The crucial roles and considerations for the installation team are:

General Contractor

The general contractor (or "GC" as it is sometimes called) is the counterpart to the architect; he holds the prime contract with the owner (or the owner's representative) for building the facility. The GC may have some "in-house" capabilities for site construction, but most of the work is managerial in nature. GCs hire out most of the specialty work to subcontractors while they coordinate, arbitrate, facilitate, and schedule all the work that needs to be completed.

Working with the general contractor

One of the GC's concerns is the successful coordination and installation of all "architecturally integrated equipment." From the AV standpoint, this includes items such as front and rear projection screens, projector, monitor and loudspeaker cluster mounts and motorized window shades. These may be designed and/or provided by the AV designer or integrator, but they are actually installed by the general contractor or one of its subcontractors. Since the general contractor has the overall responsibility for the installation, significant coordination with the AV contractor is required.

In addition, other fundamental issues need to be addressed. The first (to be discussed in the next chapter) is whether AV is sub-contracted to the GC. The second involves how the GC controls the installation schedules. Since the GC determines who works when and schedules priorities for work completion, there is an impact on the AV installation sequencing (specifically when cabling can be pulled and what areas will be ready for equipment installation). Finally, the GC may schedule site visits to monitor infrastructure installation before the AV work is completed for that stage. Some sequencing may put the AV consultant at a serious disadvantage. For instance, if a site visit is scheduled to check the wall framing and conduit installation after wallboard is put up, the AV provider cannot see if the conduit and back boxes are properly installed.

Subcontractors

There are many subcontractors to the general contractor. For the most part these are the contractor counterparts to the various trade designers previously discussed. They include contractors in charge of mechanical, electrical and plumbing and other traditional trades as well as the other specialized trades such as data/telecom and AV. Coordination of work and scheduling between these subcontractors and the other trades during the construction phase is critical to the ultimate success of any project.

Close coordination is required among the electrical contractor, the AV consultant, the AV integrator, the data/telecom installers, and the other low voltage contractors. Their installation sequencing and scheduling of infrastructure will determine when the AV designer can begin to pull cabling and start installation and testing of equipment. Back box, conduit, and cable tray locations and installation will often require review by the AV contractor and electrical consultant during construction, which also requires coordination. If any electrical outlets will serve the AV systems, they may require special treatment, such as power conditioning or isolated grounding schemes.

Other subcontractors that deserve more attention include those that provide either pre-manufactured furniture or custom-built stand-alone furniture (millwork) or built-in furniture (casework), which may need to accommodate AV equipment. If the furniture requires integration of AV equipment, then the furniture, millwork and/or casework contractor will need to know precisely what is required for the equipment integration and will most often perform the actual drilling and cutting required to accommodate the equipment. In many cases, the AV integrator may subcontract to the furniture manufacturer to provide AV-specific pieces so that the integrator can have more control over the furniture design and delivery. Close coordination is required. The end-users may also need to be involved in the design and installation process, since there may be user-provided equipment (often referred to as "owner furnished equipment" or O.F.E.) that needs to be integrated. Likewise, for large institutional users, there may be campus standards for AV furniture design and finish that need to be followed.

AV Integrator

The AV integrator (who may also be referred to as the AV contractor) installs the AV systems. Typically, this includes all the AV electronic equipment (projectors, audio and video routers, loudspeakers, amplifiers, cameras, DVD players, and other audio, video and control devices and their software programming), as well as the dedicated AV cabling, such as base band video coaxial cabling, audio line, microphone and speaker cabling, plus proprietary control system cabling. The AV contractor may also install some networking cabling (when it is required for special AV use), such as point-to-point video and audio distribution that bypasses the data/telecom patching scheme.

Due to the need for architectural integration, some equipment may need to be provided to the general contractor early in the process. In addition, the sequencing and scheduling of the AV integrator's on-site installation time is extremely important. Cable pulls must be done at the point at which there is easy access to conduit and cable trays, preferably before lay-in ceiling grids are installed.

Once delivered, sensitive electronic equipment must be kept out of dusty areas and in a secured facility. All of this requires close coordination among the AV contractor, the general contractor and the building occupants. Additional coordination is necessary with the building owner (and users) on issues such as provision of data networking, telephone and communications services, and scheduling of commissioning and user training.

AV Control System Programmer

The AV control system programmer provides the "backend" programming for the system. This programming essentially directs the equipment to perform the desired functions when the user presses buttons on a touch screen or other user interface. It can also direct the action to occur at a scheduled time or after a particular event.

This function can range from something as uncomplicated as instructing the DVD player to operate when the Play button is pressed on a touch screen to the more complex activity of monitoring the status of a projector's lamp life and sending a warning e-mail to a technician when the lamp is reaching the end of its rated lifetime.

Today, the programmer plays a much larger role in the AV integration process. With more computer network interfacing of equipment, more network-based system monitoring and control, and more control system interface design issues, the control system programming is an important piece of AV integration. In fact, many AV systems are almost useless unless the control system programming is complete and the interface design is user-friendly.

The control system programmer may be an employee of the AV integrator, an employee of the control system hardware manufacturer, an independent programmer for hire or an employee of an independent AV consultant. In rare cases, the owner provides programming for the systems.

Ideally, the programmer should be given an interface design with detailed text descriptions of the complete system functions. In addition, the programmer should coordinate computer network access with the system owner during installation and testing. This will enable the programmer to complete the programming installation and get the systems operational.

The Management Team

The owner often utilizes the services of management and consulting companies that can relieve the internal staff from the burden of managing a large project. These entities may work with the real estate, construction management or logistical aspects of the project, and are often affected by the AV impact on a new building or renovation project.

The Developer

Developers generally find land (with or without an existing building), invest in it to generate income, and prepare it for sale or lease to a potential buyer or lessee. If the building is being developed on speculation ("spec building"), they hire the architects and contractors directly. Otherwise, the developer is hired as the construction manager by an owner to assist with the site development.

The developer is concerned about the budget, schedule, and scope of the project. Because the developer is often financially invested in the project, he or she is even more sensitive to project costs and future value. The developer is well served by being given complete information on how AV fits into the project, its importance to potential owners or lessees, and the impact it will have on the design and construction process.

Construction managers should include AV in project budgets and management

The CM is concerned about schedule, scope and budget, and it's important to understand how the AV process is similar to or different from traditional building trades. AV affects a wide variety of project areas. It adds cost to the base building for infrastructure, and the AV systems have the potential to be a significant and unexpected budget line item. In fact, the AV contract may not be issued until after the base building has been contracted for. Experienced construction managers are prepared for issues related to audiovisual needs.

Construction or Program Manager

The construction manager (CM) is a separate entity hired by the owner to manage the construction process. The CM's job is to manage the schedule, scope and budget of the building for the owner. Sometimes also called the program manager (PM), the CM maintains the approved program (the document that establishes the basic parameters of the building including overall size, budget, features, quality and space configurations, allocations and adjacencies). Sometimes, the CM hires the architect and general contractor. Or, the CM merely coordinates the process while the owner holds the contracts.

It is important for the CM/PM to understand how AV fits into the project so that he or she can make the best decision about how these contracts are managed.

Building Management Agency

Some building owners may hire a building management agency to handle day-to-day operations, maintenance and leasing. Managers may be located in the buildings they manage or off-site. The management agency may be the gatekeeper for when and how contractors can work in a building. They are concerned about the operation of the building and the comfort of the existing tenants and/or disruption of their routine. As a result, they have a great deal to say about which contractors are hired and when they can work in a building.

The agency needs to be made aware of how the AV contractor may have different needs from the other trades in accessing the building, particularly at the end of the construction process.

Roles within Organizations

Most of the roles we have discussed so far are technical roles that may be performed by either an organization or an individual (and in some cases, both). There are three basic "generic" roles that are important to highlight to understand their place in the process. These are the Project Manager, the Designers and the Installers.

The Project Manager

Throughout the process, the role of project manager is common. It exists in almost every organization, but may differ from one to the next. Sometimes, the project manager is just that — a manager concerned with the sequencing, scheduling, coordinating, facilitating, and budgeting of time, materials, and actions to get the project done. Or, the project manager may also be a designer, installer, superintendent, architect, engineer or technician. In either case, the role of project manager is as important as the technical roles required to get the project designed and installed.

The project manager is concerned with getting the job done. The project manager is concerned with not only the sequencing, scheduling and budgeting issues, but also with the full breadth of coordination required with the variety of both design and installation trades that AV affects.

A project manager's role has many facets, but core responsibilities include:
- Development and maintenance of a project schedule and table of responsibilities.
- Timely and consistent communications to and from all members of the project team; the PM should be the single point of contact for his or her team.
- Supervision of all project resources with respect to schedule and accountability to the project (includes people, equipment, rooms, etc.).
- Oversight and influence over adherence to the defined project schedule and negotiation with other teams if the schedule needs adjustment.
- Assurance that the teams work in harmony with all others on the project.

The Designer, Consultant and Engineer

The designer has the technical expertise to assess the user's needs and translate those needs into documents that convey the design intent to the installers, usually in the form of drawings and specifications. Designers exist in almost all design trade organizations. The designer may work with a separate project manager or may serve as the project manager for his or her organization.

Though some designers may operate without formal certification, they should still be thoroughly knowledgeable in their field. For AV designers, holding a certification such as InfoComm's specialized Design Certified Technology Specialist (CTS-D) is recommended. Many non-AV designers, such as architects and mechanical, electrical and structural engineers are required by law to be certified and licensed because of the life safety issue associated with their designs.

The Installer

Installers are responsible for interpreting the design intent depicted in the contract documents created by the designer and assembling the item or system in the manner described. The installer may need to be certified and/or licensed by law, and may or may not be part of a labor union which may affect how some projects are constructed in certain localities. Like designers, installers often work with a project manager who coordinates and directs their work, or their roles could be combined.

The installer is often the source of questions and coordination throughout the process when clarifications, conflicts or design questions are addressed. Good communication among the AV designer, AV installer and installers of related trades is important in making the installation process both productive and smooth.

Others in the Design and Construction Process

In addition to the teams and individuals noted so far, there are a few other participants in the process that do not fall neatly under one of the categories covered. They are:

Communications Service Providers

Communications companies provide services, such as telephone (or "POTS," plain old telephone service), digital telephone and data services (ISDN, T-1, E-1), and Internet access. Generally, the owner arranges these services, many of which are integral to the AV systems operations, particularly where audio- and video-conferencing are part of the AV system. Internet access is important for users at their workstations and for the AV system to be remotely monitored, scheduled and diagnosed off-site. Coordination and scheduling of the provision and activation of these services is crucial during the testing and commissioning phase of the project.

Commissioning Agents

A growing trend with building owners is to hire third-party commissioning agents to test and verify that the general building systems are operating properly and meeting the specifications of the construction contract. These agents may be involved in reviewing drawings and specifications during the design phase, reviewing systems installation during the construction phase, and testing/training at the end of the installation process. As of the publication of this manual, relatively few commissioning agents have the expertise to address AV systems adequately. If they are commissioned to do so and have the expertise, the AV consultant and contractor will need to convey the design intent, documentation reviews, and final testing/training to the agent.

Code Officials, Inspectors and Local Authorities

All new buildings and renovations are subject to local, state, and federal codes and regulations that have an impact on the bricks-and-mortar parts of the building, ranging from electrical systems to handicapped access. There are usually three stages of involvement for officials who administer these codes:

1. Design Phase Submittals
   The architect's plans are submitted to the officials for review and approval before the project can be bid or constructed.
2. Construction Phase Inspections
   The installed elements need to be inspected by the officials and signed off before construction is complete.
3. Occupancy Approval
   At the end of the project (particularly for new buildings and large renovations), an inspection is required for the issuance of a certificate of occupancy to allow the owner and/or end-users to move in.

Code officials look for items that have to do with life safety and handicapped access. This covers virtually every aspect of the building, from corridor and stair widths to emergency lighting, and from structural designs to electrical system grounding. Some of these are AV-related, such as audio muting associated with a fire alarm system and power system grounding. These are issues that will have to be addressed either by the AV contractor or by the electrical or other contractor.

CHAPTER 2
Understanding the Process

CHAPTER TWO

This "roadmap" chapter offers a logical guide to deciding on an appropriate process for making the AV project a reality.

What parameters are involved in deciding which process to use? What options are available? And what impact will your choice of options have on the whole process? Specific tools are recommended for assessing the proposed AV project and the issues that have a direct impact on the ultimate result. These issues revolve around:

- Schedule
- Budget
- Policy or legal issues
- The owner's structure and expertise

Specifically, this chapter details the two most common processes — Independent Consultant Led Design-Bid-Build and Integrator Led Design-Build — for AV project delivery, along with the pros, cons and risks of each.

Also covered are the less commonly used hybrid processes: Consultant Led Design-Build, OFE/Integrator Installed, OFE/Owner Installed, and a newer configuration, Consultant/Integrator Team Design-Build.

Determining which process to use is the single most important decision in AV design and integration. The right choice comes from reviewing the parameters, having knowledge of the available options, and understanding what impact those options will have on the entire process. Parameters that impact the selection include:

- Nature of the project — What is the project timing and type?
- Financial support — How will the project be funded and sustained?
- Legal and policy issues — What impact do they have on the process?
- Owner's structure — What type of organization is it?
- Owner's expertise — What is their level of AV expertise?

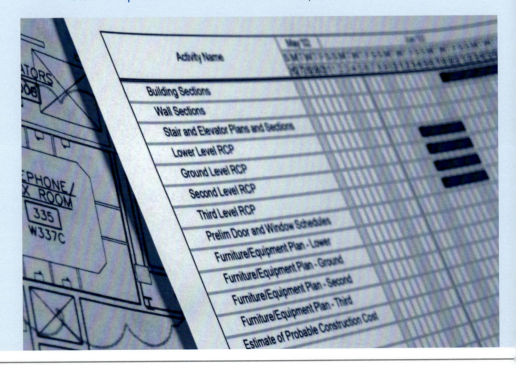

Assessing Project Characteristics

The five basic parameters that play a key role is assessing which method is best suited for a particular project are the project's schedule, size, type, complexity and procurement requirements. Understanding how each affects the decision will allow for thorough planning and delivery option decisions.

Predictability of outcome is critical, given the complexity and importance of these communication systems to the owner's business. Often the owner chooses a process based on a balance of project definition and anticipated project risk. A longer design process is used to help define project scope, investigate and document solutions, and thus limit financial project risk to the owner during installation.

With system innovation, new technology and usage of unique combinations of technologies or software, the end result is not assured or predictable, the project has more risk and the owner will seek to protect himself with a higher level of investigation and quality assurance prior to contracting. (This may result in mock-ups of systems prior to full design and process selection.)

If the project is perceived to have sufficient definition or is fairly standard in the technology used, a simpler shorter process, involving less upfront work and cost, may be used with limited risk.

The opposite argument also holds that the owner will look to one entity to assume full responsibility for the solution and the risk given the complexity of the systems. In this case, the owner will look to the most "responsible" process/party to both design and deliver the solution.

The key to either process is the owner's planning and management of the risk. Often it is the owner's expectations that create the risk. Tight scheduling can also heighten the risk involved for all parties.

Project Schedule
The project schedule will have a direct impact on many decisions, and can be the determining factor in choosing which procurement process to use. A shorter project schedule may not allow for the full design-bid-build process, in which case a design-build process led by an AV integrator may be the better solution. A longer project schedule may be better suited to a design-bid process that is led by a separate independent consultant.

Project Size
The size of the project is an important parameter in deciding which process to use. For example, a large AV integration project for a new building that allows a longer schedule may be better suited for an independent consultant-led design-bid-build process. A small project with a project schedule of less than six months is generally better suited for an integrator-led design-build process. Other factors may determine which is best for "in-between" projects.

Project Type
An AV project may take a number of different forms. It may be a stand-alone upgrade or part of a new building. It may be a part of a renovation or a retrofit of a previously technology-free space. It may even be a maintenance project, where only selected items are upgraded, or a system upgrade, where all, or part, of an existing system is replaced with newer and/or more usable equipment. The project type is another important factor, along with size and schedule, that should be carefully considered in determining the process to use.

Best Practice

Put AV systems in the capital budget

Always determine the appropriate category for the AV budget at the beginning of the project budgeting process.

Normally, the AV systems in a large installation project should be identified as capital budget (the same as other traditional building trades). It should not be considered FF&E. The operations, upgrades and augmentation after installation would be appropriately considered operations and maintenance budgets. The only AV equipment that might constitute FF&E would be portable equipment that is not associated with permanently integrated systems.

Project Complexity

AV projects range from simple one- or two-room installations to very complex, expensive systems. They can consist of many smaller, less complex systems or a single large, sophisticated one. Each of these conditions, in tandem with the other parameters cited in this chapter, will play a part in determining what delivery process or options are best for a particular AV project. A complex project with a short time-frame may call for the involvement of an independent consultant to address the peculiarities of a particular vertical market (e.g. museums, or retail spaces) or special infrastructure issues; a less complex project of the same size and schedule may be a better candidate for an integrator design-build solution.

Procurement Requirements

Regardless of project schedule, size, type or complexity, there are many instances when procurement requirements that exist within an organization or government entity will dictate the project process. If local, state or federal law (or institutional rules) requires competitive bidding on AV equipment and integration services, then a consultant-led design-bid-build process may be the only viable path for an AV project to follow.

Funding AV Systems Installation and Support

For any AV project, money is a make-or-break issue. There are three important aspects of funding that have a direct impact on both the project process and the system design/installation:

1. The AV System Budget

The AV budget can be a significant budget line item in the initial facility budgeting process. In fact, in many of today's high-tech building projects, the AV sub-contract represents the largest sub-contract in the job. It should be evaluated with the help of qualified professionals who can provide reasonable estimates for the initial anticipated levels of AV technology desired for a new room or building. If the budget is clearly insufficient, then either the systems must be reduced in scope or additional funds must be found. It is always better to determine the relationship between available funds and system requirements early in the project.

In addition, there should be a determination of the source of the budget. For example, is the AV budget coming out of capital funds? Is it part of the base building costs, similar to the electrical or fire alarm system? Is it part of Furniture, Fixtures and Equipment (FF&E)? Or, is the money coming out of an operations budget?

2. Funding Options

The funding of full-fledged building projects, renovation project upgrades, one-off projects, and repair projects are handled differently from organization to organization. (In fact, each situation may be handled differently even within the same organization.) Building projects may be funded from a private, internal corporate building capital fund, or publicly funded with approvals tied to a legislative body or a board of regents.

Correct appropriations should be made as early as possible if the building is to include significant AV technology. Once a budget is set and approved, there is often no opportunity to request the necessary funding for the AV system. While operations, maintenance and upgrade projects may have more budget flexibility, early appropriation is still the best course of action.

3. Technical Support Funding

The third aspect of funding that can have an impact on the system design, installation and operation is technical support staffing. In reality, many projects involving the installation of a multi-million dollar AV system are completed, only to operate under their full potential because of unprepared or understaffed internal AV crews.

In most organizations that have enhanced their AV systems over the years, their internal expertise has generally grown as well. These organizations probably have one or more staff members who are well versed in AV systems design, procurement, installation and operation. Many other organizations, unfortunately, have very limited experience with systems. If the plan is to move quickly from a few unsophisticated AV systems to a large system or multiple smaller systems, the current staff may be unequipped to handle the complexity and scope of the new AV systems. The organization also may not have the agility to adapt as quickly as necessary to the new AV landscape.

This potential disparity clearly raises two issues, which should be addressed in the AV project process:

Accessing Additional Training

New systems bring new technology and equipment that are unfamiliar to an owner's staff. The current AV technical staff may need additional training to be able to handle the new help desk requirements, operations concerns, and maintenance issues. Additional funding to train is therefore required, because this type of training often goes beyond the basic user training provided by the AV design and installation team. The staff may need technical training provided by individual equipment manufacturers, AV industry associations[12] or third-party providers.

Hiring Additional Staff

Even with additional training, the existing staff may not be able to manage the advanced systems. In this case, more staff (as well as more training) may be necessary. Alternatives to adding full time personnel are outsourcing (on an ad-hoc or per-event basis), contracting long-term, contracting with full-time staff either privately or from an AV provider, or using the services of an independent consultant.

Get educated about AV

If an organization is considering an AV project, it is imperative to get up to speed quickly. That requires research and education. Understanding the players, their roles, and the available options will improve the chances of greater AV project success. The best way to learn quickly is to become active in AV industry trade organization, such as the International Communications Industries Association (InfoComm), which offers numerous vehicles for learning and peer-networking. The exchange of ideas provides "real world" perspective and confidence.

Legal and Policy Issues

Large private companies, public organizations, and government agencies have policies or legally established procurement procedures that determine which procurement process to use for AV systems.

Generally, these policies work well for traditional building projects, but often are not well suited for AV systems. For example, the requirement to bundle construction and AV systems bids can be restrictive and cause AV technology to be bid too far in advance of installation, creating the need for potentially expensive change orders later in the process.

If the policy is unworkable, it may be important to seek a "situational" policy change — one that allows for a revision to the normal policy to maintain the integrity of the AV bidding process.

---

[12] Extensive AV and information communications training is offered through the International Communications Industries Association (InfoComm) in a number of formats, including online courses and in-depth and hands-on training. In addition, at its annual conference and exhibition, InfoComm, numerous courses are available including product-specific courses from manufacturers. Check www.infocomm.org for additional details.

Owner and End-User Expertise

The owner is the entity buying the system and is frequently also the end-user. The owner and end-user require a systems delivery process that meets their needs and fits in with their operations. An important consideration in determining how to implement a new or renovated AV system is how much AV expertise the owner has in-house. If the owner has experienced AV professionals on staff, then there may be more flexibility in choosing who to provide AV services for a particular project. If the owner has little or no internal expertise, this may lead to a consultant-led process to provide

Figure 6: Large Construction Project: Typical Schedule

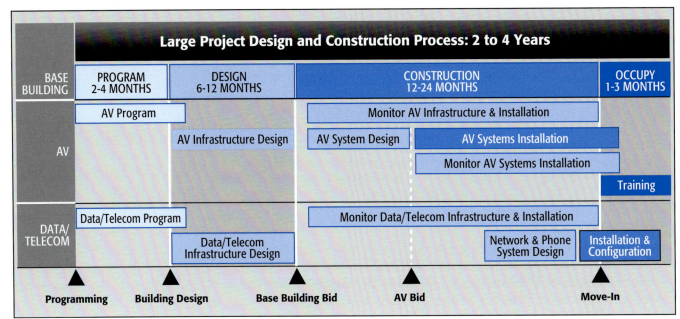

independent technical representation for the owner. If, however, the owner has a good relationship and experience with an integrator, an integrator-led process may be equally suited.

Examining the Processes

Two basic and common processes exist for AV project delivery:

- Independent Consultant-Led Design-Bid-Build
- Integrator-Led Design-Build

There are four less common variations on the above options including one that is completely owner-based:

- Consultant-Led Design-Build
- Consultant/Integrator Team Design-Build
- Owner-Furnished Equipment/Integrator Installed
- Owner-Furnished Equipment/Owner Installed

Choosing which process to use depends on all the variables discussed previously. Each process has its positive aspects (pros), negative aspects (cons) and some significant concerns (risks) that need to be reviewed.

Project Scheduling: Where Does the AV Project Fit In?

A good project schedule for a large, technology-enhanced building will reflect the critical milestones that occur in a full-blown AV project. In a smaller building, many of the milestones will have already been met and others may overlap. Regardless of the size of the project, the basic procedure is the same.

Figure 6 shows the overall schedule for a new building that includes significant AV and IT technology integration. In the early stages of the project, the schedule includes architectural and technology programming (or needs analysis) and budgeting.

Figure 7: Smaller Construction Project: Typical Schedule

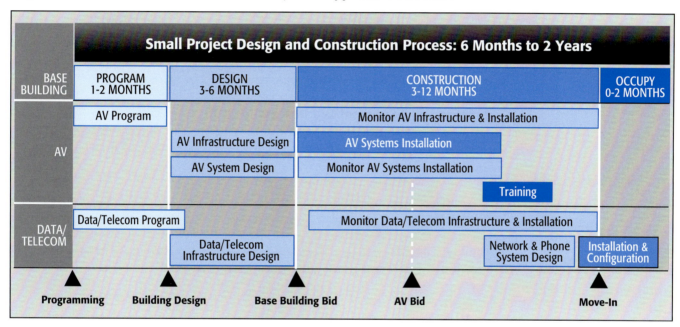

AV and Base Building Design and Construction: Parallel Tracks

In AV design and integration, there are two basic tracks of design that occur before construction and integration — infrastructure design for the base building and electronics design. Though intimately related, these two design and installation tracks run separately in many ways.

The base building track involves the acoustics, lighting, conduit, data, power and space requirements that need to be incorporated into the building design and will generally be installed by the appropriate trade contractor.

The AV electronics track includes the AV equipment (audio, video, AV control and AV wiring) that is supported by the infrastructure systems. The AV integrator usually installs the AV systems, although in some cases, the owner's staff may install the equipment.

For a new large facility, the base building infrastructure is designed during the architect's and base building engineer's design phases as shown in Figure 6. Due to the length of the design and construction process, the bulk of the electronics design is often completed after the base building is contracted and construction is underway.

For a smaller and/or short-term project process (Figure 7), the schedule elements are basically the same, but the infrastructure and electronics designs will often overlap so that both are in full swing at the same time for both the design and installation phases.

This project schedule will include the same details regardless of the AV delivery process chosen.

In a large project, these activities can take from two to four years or even longer. The AV electronics are generally not designed until later in the design and construction process, while the infrastructure for the systems is designed along with the rest of the building as indicated in Figure 6. If the electronics are designed too early, many model numbers may not be available by the time the system is purchased and installed. In addition, the technologies may have changed (a likely scenario in light of today's rapid technological advances) and new approaches may be much more desirable.

For smaller projects, the schedules tighten up and the AV design and installation schedules may more closely follow the base building schedule as shown in Figure 7.

Assessing the Traditional AV Project Delivery Processes

As the AV trade has evolved, two primary methods have developed that are in common use today. They involve two key AV providers: the independent AV consultant and the AV integrator. One method mirrors the traditional Architect-General Contractor based design-bid-build process that is often used for new and large-scale renovation projects. The other is a design-build option that is similar to the design-build concept sometimes used for a building project.

Independent Consultant-Led Design-Bid-Build

The consultant-led design-bid-build process involves the use of an independent consultant who operates similar to, and in conjunction with, the other building design consultants on the architectural team — during both the design and construction phases.

The independent consultant develops some, or all, of the infrastructure and electronic systems designs. The AV consultant monitors and administers the installation phase (performing as the owner's technical representative) just as the mechanical and electrical engineers do for their systems. At the end of the process, the consultant works with the owner and AV integrator to provide final commissioning and training on the installed systems.

Contracts

The consultant is usually contracted to the architect along with the other building design consultants. The responsibility for the design rests with the AV consultant and the architect who contracts the consultant. On occasion, the consultant may be contracted to the owner directly, but this is not the norm. The AV integrator is usually contracted to the general contractor and, in some cases, is subcontracted to the electrical contractor. (The latter is not recommended because of the level of interaction required among the AV consultant, the AV contractor, the general contractor and the owner.) A typical contract configuration is shown in Figure 8.

At times, it is advantageous for the AV integrator to be contracted directly to the owner. The biggest advantage is that the owner will be relieved of paying the subcontractor "markup" that is usually applied by the general contractor to any subcontractors' costs. Since the bulk of the AV contractor's work is done later in the construction process (much of it just before and after occupancy of the building), there is a shorter timeframe for detailed coordination on the part of the GC and the AV contractor as compared to the other trades.

When To Use It

This method is most commonly used in larger new building projects. The AV budget (not the building budget) and the project schedule are the key factors. If the project is a new building that has a design and construction schedule of more than two years, then the consultant-led design-bid-build method is usually the best choice. New buildings with AV systems that cost approximately $750,000 or more are also the best candidates. This process will not be possible if the project schedule is too short; there may not be enough time to complete all of the process steps sequentially.

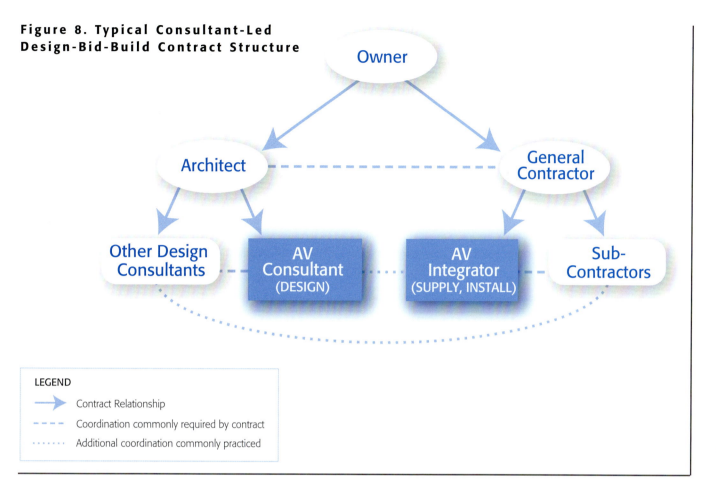

Figure 8. Typical Consultant-Led Design-Bid-Build Contract Structure

LEGEND
→ Contract Relationship
---- Coordination commonly required by contract
······ Additional coordination commonly practiced

Pros

- Good for longer-term projects. Consultant's business model is built on extended project cycles.
- Consultants are generally more experienced in working with the architectural design team and often have the various infrastructure specialty fields (acoustics, lighting, data/telecom, and others) covered in-house.
- The owner has independent technical representation for AV and infrastructure issues similar to the architect's role.
- Continuity provided from program phase through commissioning and training for long term projects.

Cons

- Not appropriate for smaller projects or projects with short schedules.
- Depending on the project contracts, there can be more than one point of responsibility for the AV system design.

Risks

- Hiring under-qualified or inexperienced providers. The consultant, integrator, architectural designers and construction team must work together to make the project a success. Any weak links will put a strain on the process and the quality of the work results.
- This process works best using independent consultants. Hiring an AV integrator as the consultant may create conflicts with the independent consultant if the integrator bids on the job they design.

Integrator-Led Design-Build

This method of procuring AV systems involves using the AV integrator for both the design and construction phases of the project.

In this option, the integrator is required to provide all the base building design items (acoustics, lighting, power, data and conduit), as well as to design and install the systems. On occasion, the integrator may subcontract to an independent AV consultant to assist on a large and/or complicated project to augment the integrator's expertise.

This process is built on the premise that the same integrator does the design and the installation. If the integrator is contracted for the design phases separately from the installation part of the work, the hiring organization may decide to "shop" the installation. This action could require the integrator to prepare biddable drawings and specifications. It is important to resolve this issue early in the process to prevent a change mid-stream. This action would only be acceptable if there are performance problems with the first integrator and a change is necessary.

Figure 9: Integrator-Led Design-Build Contract Structure

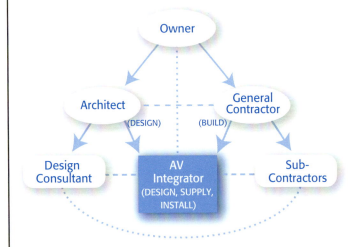

For this method, the integrator may be contracted to the architect or the GC.

LEGEND
→ Contract Relationship
---- Coordination commonly required by contract
...... Additional coordination commonly practiced

Contracts
Integrators are typically contracted to the architect or to the owner for the design phase; they are contracted to the general contractor or to the owner for the installation phase. As always, subcontracting to the electrical or other subcontractors should be avoided.

When to Use It
The design-build process is most commonly used for smaller projects (less than $100,000 in AV systems costs). This also includes small AV projects that will be installed in larger architectural jobs. The design-build process does require that the AV integrator be prepared to work on building infrastructure issues a year or two in advance of the actual AV installation.

This process can work for larger projects if the contractor has experience in a vertical market, such as health care or aquarium facilities. It is also effective for more common vertical market projects, such as corporate, government, and education, if the integrator has the expertise and a project management structure that can sustain longer project timelines. Integrators need to have the ability to provide consulting-only services well in advance of AV system sales.

This process is generally inappropriate for large projects (AV system budgets significantly over $1M) except under very special circumstances. In addition, any project where the design and installation phases last longer than two years would be better suited for one of the consultant-led options.

Pros
- Provides one source of responsibility for AV and infrastructure design and installation.
- Final contract costs can be established earlier in the project.
- Especially appropriate for shorter term projects.
- Extensive day-to-day field experience with products and technologies.

Cons
- More owner involvement and responsibility required compared to consultant-led models.
- AV systems designs may be influenced by equipment sales incentives, which raises questions of compatibility with the system requirements.

Risks
- The owner, architect or general contractor must be able to evaluate potential integrators through requests for proposals and oversee their work, or the quality of the project may be compromised.
- Most AV integrators are set up for shorter-term projects and may not have the expertise or time to handle the full range of infrastructure design required during the base building design phases of larger projects.
- Sometimes the owner or general contractor may choose to "shop" the design provided by one integrator to others during a design-build process. This practice undermines the process and should be avoided during the course of a typical design-build process.

Choosing Design-Bid or Design-Build Based on AV Budget and Schedule
Figure 10 on the next page can be used to make an initial determination about which of the two basic methods would be best for a particular combination of construction schedule and AV systems budget. In using this chart, it is important that the AV budget figure being used is realistic and appropriate for the systems being considered.

Process Variations for Consideration
Beyond the two more common methods just described, there are less common methods that may be used to deliver AV systems under some circumstances.

Consultant-Led Design-Build
This option is a hybrid of the previous two methods. An independent consultant is hired to provide initial services, such as needs analysis, but is not responsible for developing the detailed systems design. The integrator is usually hired separately.

Figure 10: Method Selection Chart

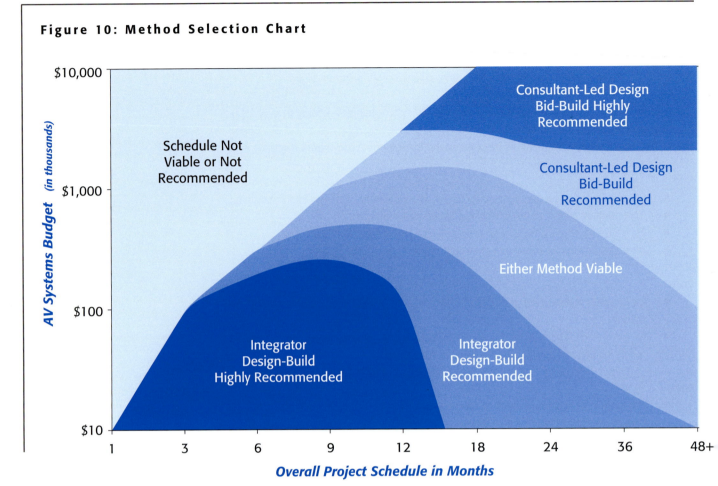

This graph provides guidance in choosing between the basic delivery methods for delivering an AV system. First, locate the approximate value of the anticipated AV systems to be contracted on the left side of the chart. Then locate the approximate overall schedule of the facility design and construction (from initial design to end-user occupancy) along the bottom of the chart. The area that falls under the intersection of these two provides a guide to the method that would be recommended for that combination of budget and schedule.

Under this method, the consultant still acts as the owner's technical representative, and may provide some facility infrastructure design or recommendations, submittal and site monitoring during construction. The integrator provides most of the systems designs and performs the installation. The consultant might also assist with final testing, commissioning, and training.

Contracts

The independent AV consultant would normally be hired directly by the architect or the owner. The AV integrator is hired separately and may be contracted to the owner directly or to a general contractor. Subcontracts to the electrical contractor or other subcontractors is not recommended.

It is rare that the consultant would be hired and subcontracted to the integrator. The consultant could be hired by the integrator to assist and provide the same scope of services, but direct contracts for the consultant and integrator separately is recommended as the best approach.

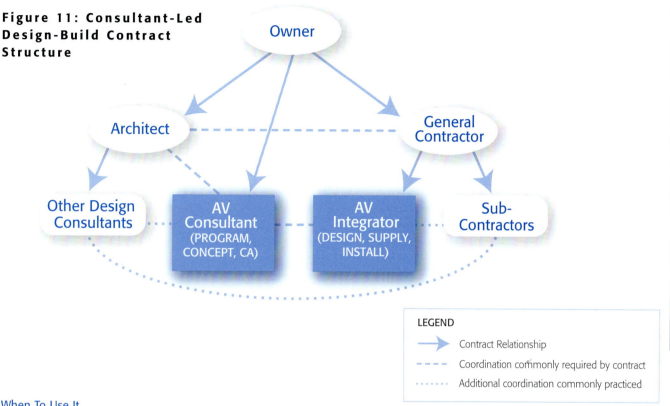

Figure 11: Consultant-Led Design-Build Contract Structure

When To Use It

The process could be used for a smaller project when a complete scope of consultant services (under a design-bid-build scenario) is not cost effective or desired. Other circumstances that may lead to using this method in lieu of the usual integrator-led design-build are:

- The desire of an owner to have third-party technical and/or managerial assistance because it is unavailable in-house.
- The system is unusual or complicated and would benefit from particular applicable experience beyond the integrator's portfolio.
- There are special facility conditions, such as lighting, acoustics or other infrastructure issues that require expertise beyond the AV integrator's capabilities.

Larger projects that require significant facility design or have complicated systems with very short project schedules are also appropriate for this process. The consultant can provide technical representation and project/program management for the owner, while the integrator is on board concurrently to develop the detailed designs and provide installation to meet the tighter schedule.

Pros

- This method is very flexible and can be used for almost any project. The consultant's involvement can be limited to a reasonable amount of "looking over the shoulder" of the integrator in smaller projects or expanded to include some infrastructure design, shop drawing submittal review, installation monitoring and system commissioning.
- Allows consultant to provide AV technical representation for the owner within a design-build environment.
- Can allow for more infrastructure design expertise than traditional integrator design-build process.
- Enhanced quality control for short term, fast track projects.

Cons

- While it may be a useful alternative, some consultants and integrators are uncomfortable with this arrangement, and the scopes of work for each may be more challenging to define.
- Not as appropriate for very large projects with longer timeframes and more significant infrastructure design.

Risks

- This process, like the others, depends heavily on the qualifications of the AV providers. Knowing how to select the appropriate AV providers is important to avoid problems with the design and installation quality.

Consultant/Integrator Team Design-Build Approach

This team approach is very similar to the consultant-led design-build option; in this case the consultant and integrator form a joint venture that is usually contracted as a single entity. Architects and general contractors often team up this way for design-build projects and this approach is likely to be more common for AV projects in the future. Contracting for the joint venture is similar to the design-build model, but the roles and responsibilities are more akin to the consultant-led design-build option.

OFE/Integrator Installed

Under this scenario, the owner buys equipment from one source and uses another source to integrate it.

While in this case an AV integrator could provide infrastructure and systems designs, the list of equipment provided may or may not work. This method works best when the owner has the expertise in-house to design and manage the AV systems that are procured.

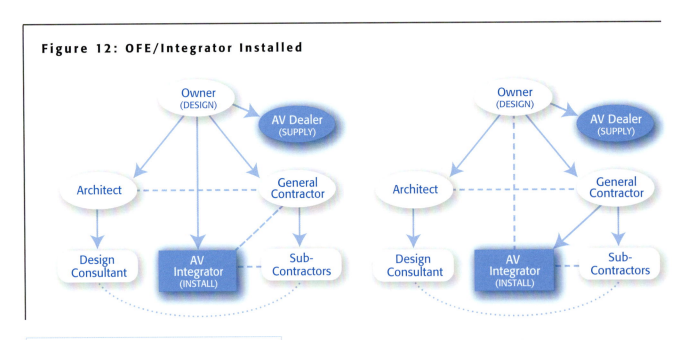

Figure 12: OFE/Integrator Installed

For this method, two common variations are shown indicating that the integrator may be contracted to the GC or directly to the owner.

Contracts

In this situation, the owner buys equipment from either an AV integrator or a "box-sale" vendor that does not provide installation services (an AV dealer as defined earlier). The owner then contracts the integrator directly, (or subcontracts the integrator through the GC), to integrate the purchased equipment into the facility.

When to use it

This process is most appropriate when it is a small part of a project that falls under one of the other processes described previously. For example, an owner may be developing a new room or facility that includes new design and equipment, but he or she wants to integrate previously used equipment into the new system.

The process could be used for the whole project if the owner has reasons to purchase all of the new equipment independently. These could include budget restrictions (e.g. money can be spent on equipment, but not labor), tax advantages, and/or the requirement to spend last year's budget before the end of the organization's fiscal year. The owner will then wait to purchase the integration.

If an owner wants to move an existing system from one facility to another, he will often furnish the existing system and hire an integrator to re-install it. While this sounds feasible, there may be differences between the new facility and the old one that will require new equipment and/or modifications to the design.

Pros

- Allows owner to have budget flexibility by reusing existing equipment.

Cons

- Equipment and system damage protection and warranties are jeopardized because the integrator installing the equipment did not sell it to the owner.

Risks

- System and infrastructure design and facility functionality may suffer since purchases are often made before there is an adequately developed system and facility design.

OFE/Owner Installed

Large organizations with many AV systems often provide internal AV system design and installation, as well as operations, maintenance and management. These owners may elect to purchase and install the equipment themselves under some circumstances.

In these cases, the owner performs a needs analysis internally, develops the AV systems and infrastructure designs, purchases the equipment from an AV vendor, and uses in-house staff to install the systems. Under this option, it is still best practice to follow all of the steps noted in the phases described later in this book.

Contracts

The only external contracts are between the owner and the AV dealer. In large organizations, corporations or government agencies, however, there may be an internal "contracting" process from one department, or profit center, to another.

CSI MasterFormat and contracting structure

The CSI MasterFormat 04 created a separate Division 27 – Communications that includes specifications sections for data/telecom and AV. CSI has long maintained that the MasterFormat divisions are not intended to create trade jurisdictions. This should be considered as MasterFormat 04 comes into use. AV should be contracted separately from data/telecom unless the contractor has acceptable experience and capabilities for both areas of expertise.

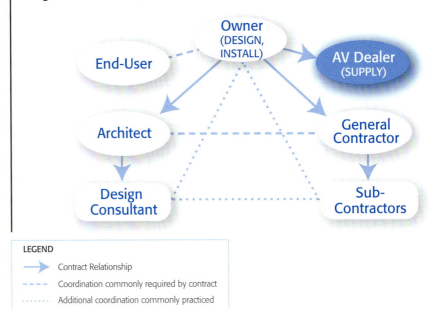

Figure 13: OFE/Owner Installed Contract Structure

LEGEND
- Contract Relationship
- Coordination commonly required by contract
- Additional coordination commonly practiced

When To Use It
This method should be used only when the owner has all the necessary internal AV management, design, and installation staff to undertake the project. In many cases, only projects of a certain size are handled internally. Larger projects would require that an external professional AV provider be brought in.

Pros
- Can be more cost effective with more control over the design and installation process.
- Can be conducive to good internal coordination between the designers, installers and end-user organizations.
- If the organization can maintain the internal staff, this can be a cost-effective AV management strategy for AV-rich organizations.

Cons
- The owner must manage the equipment and system warranties internally.
- In maintaining an internal organization, the owner will have to deal with staffing issues, e.g., qualifications, competitive compensation and retention.

Risks
- Staff may not have the broad-based expertise or time for a complete AV facility infrastructure design (lighting, acoustics, space planning, etc.), and the internal team may need to be augmented with specialty consultants.

Choosing a Process
The decision to use one method over another may not be a clear one. The owner and architect are usually charged with making a choice of the method or the AV providers or both, and they need to evaluate the options. Project schedules and AV budgets are the most important parameters with which to begin. Then an evaluation must be made based on the circumstances and the pros and cons for each alternative to find the best fit.

CHAPTER 3
Selecting and Contracting the Project Team

CHAPTER THREE

Selecting the right project team to successfully deliver results based on the unique aspects of each project is a critical step and needs careful consideration.

This chapter defines in detail the approaches to determining the right AV provider, how to assess a provider's qualifications and how to bid the project. Special emphasis is given to preparing a level playing field so that all providers under consideration have an equal opportunity to present their capabilities.

An overview of the selection process is provided, as well as explanations on how to make choices in selection options, the specific techniques for bid situations, how to prepare good bid documents, and how to conduct the bid process.

A major portion of the bidding process is the contract preparation. This chapter covers the strategies for contracting, various contract configurations based on what process is being employed, and, specifically, how contracts should be handled.

There is also an objective discussion about software ownership and licensing. This issue will become increasingly important as software development and configuration become more essential to pro-AV system design and installation. These discussions and suggestions lay the groundwork for eventual resolution.

Considerations in Building the Project Team

Delivering an AV system involves the entire project team, almost always with the help of an AV consultant, an integrator, or both. To build the best project team, consider these five factors that will guide you in selecting and contracting with these providers:

1. Project Size, Type and Complexity
The project size – in dollars and in scope – is a key factor in choosing a method for an AV project. This parameter, along with the type, vertical market and complexity, will help determine the most qualified AV provider for the project.

2. Project Timeframe
The selection of the consultant and/or integrator under a consultant-led process might be made at the same time or years apart. In this case, two selection processes are required – one for the consultant and one for the integrator. The consultant would be selected and hired for the program, design and bid stages. The integrator would then be selected for the installation. (In a fast track process or large project, the integrator could be selected before the normal bid period).

3. Contract Structure
The two considerations above are key factors in selecting the delivery method for engaging an AV provider, as shown in Figure 10. After selecting the method, there are options to consider in determining how the AV providers will be contracted. This step is primarily influenced by anticipated or existing contract relationships among the owner, architect, general contractor and subcontractors. The owner or architect often enters into the initial contract with the AV provider, and both should be involved in the selection process.

4. Clarity of Scope of Work

Assuming the owner's AV needs are based on a quantifiable scope of work, the complete design or design-build contract can be established immediately. Alternatively, an initial contract with an AV provider may be necessary to establish or verify the systems that are needed. After that analysis is finished, the selection and contracting process can continue.

5. Basis of AV Provider Selection

For any design and construction services, the selection of a provider may be based on qualifications, cost, or a combination of the two. A project award based on value (a best practice in the AV industry) combines the best of both.

AV Provider Selection Strategies

Once a method is chosen for a project, the AV provider is selected. There is a range of options for making the evaluation and selection, a process that is generally led by either the architect or the owner, depending on which method is used and how the AV providers are to be contracted (see Chapter 2 for typical contract hierarchies). Under any of the delivery methods, the selection process should be a team effort.

The selection team should include representatives of the project stakeholders. The team makeup can range in size. For smaller projects the team might include an owner's representative along with the architect; for larger, technology-centered facilities, the team might also include representation from both technical and non-technical end-users.

As in any selection process, the goal is to match the potential providers with the project. In AV integration projects, it is important to identify the qualifications of consultants and integrators so that they can be matched appropriately to the project. The following table lists some overall qualifications that should be evaluated when selecting a provider.

Based On:	Look For:
Project scope and size	<ul><li>Experience with projects of a similar scope and size</li><li>Size and structure of firm</li><li>Financial stability</li><li>Project management capabilities</li><li>Proven ability to meet schedules</li><li>Company credentials/awards</li><li>Current workload</li></ul>
Project type	<ul><li>Experience in a similar vertical market</li><li>Employee experience</li><li>Staffing qualifications including industry certifications</li></ul>

These qualifications will be explored in more detail after a review of the basic options for selecting a provider: qualifications-based and fee-based. It is best to use a qualifications-based selection process whenever possible. The procedure varies, depending on whether an independent consultant or an integrator is being selected.

Steps in a Conventional Qualifications-Based Selection Method

This method[13] is recommended when selecting either an independent consultant for a design-bid-build process or an integrator for a design-build process.

1. Create a request for qualifications (RFQ) that describes the project and requests specific information from the potential service providers.
2. Target the potential providers for the project by using general formats (including advertisements) and pre-selected lists of potential firms to invite to the process.
3. Send out the RFQ to the prospective firms.
4. Review the responses and create a "short list" of three to five firms to interview.
5. Interview the short list firms, using a scoring system to rank the interviewees.
6. Negotiate with the highest-ranking firm to determine the fee for the project.
7. If an acceptable fee and scope of work cannot be negotiated, move on to the second-ranked firm and so on until negotiations are successful.

Steps in Qualifications-Based, "Two-Envelope" Bid Method

While this method is qualifications-driven, it takes into account the fee as a determining factor in the provider selection. The two-envelope method is recommended and best suited for selecting integrators for an installation-only contract.

To solicit comparable fees from multiple bidders, the project description must be detailed, clear, and complete. This is best developed under the design-bid-build method of AV project delivery. The qualifications and RFP numerical bid responses are provided in separate, sealed envelopes. The qualifications are opened first and reviewed separately before opening the bid envelope.

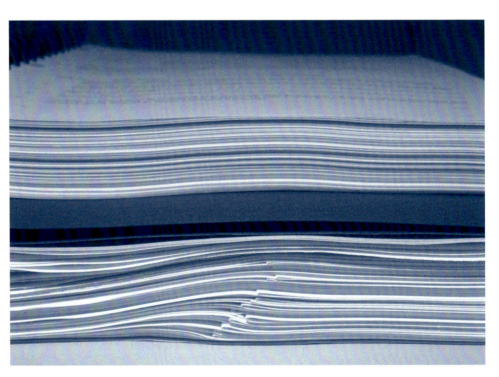

[13] The Brooks Act, enacted in 1972, requires this type of procedure in the selection and contracting of architectural and engineering services for most federal buildings. Text from the Brooks Act is included in the appendix.

The steps for this process are:
1. Create RFQ and request for proposal (RFP) describing the project and requesting the same information from potential service providers.
2. Target the potential providers using general advertisements or a pre-selected list of potential firms.
3. Send out the RFQ/RFP to the prospective firms.
4. Review the qualifications response *without opening the proposal envelope* and create a "short list" of three to five firms to interview.
5. Interview the short list firms, using a scoring system to rank the interviewees.
6. Open the proposal envelope for the top-ranked firm. If the fee is in budget range, this firm wins the bid.
7. If the fee is too high, the next-ranked firm's proposal is reviewed; this process is repeated until a proposal is opened that falls within budget.

In some cases, variations on this process may be used. A common one is to evaluate the fees and the qualifications simultaneously, using a scoring mechanism that includes a score for the fee responses to determine the highest-ranking firm.

For a process that includes fee as part of the basis, the two-envelope option is the method of choice.

Dealing with Fee-Based Selections
For projects in which only equipment is being purchased, it may be appropriate to use a fees-only evaluation to select an AV dealer — assuming the dealer is reputable and there is little risk that products will not be delivered as ordered. Selecting a consultant or integrator based solely on fee is more problematic.

Suggestions for Fee-Based Design and Consulting Service Selection
"Apples-to-apples" fee comparisons for design and consulting services are often difficult to obtain unless the RFP contains an appropriate and detailed scope of work. If the RFP is not sufficiently detailed, then the consultants' services proposed by each respondent may differ in the deliverables provided, the level of construction monitoring provided, and the amount of training and travel included in the fees.

Suggestions for Fee-Based Integration-Only Selection
Using fee alone as the basis for choosing integration-only services is not recommended. Selecting the best AV integrator for the job is important with any project, and this is particularly true for large and complex AV projects where the stakes can be especially high. Yet more than a few fee-based selection processes have resulted in hiring firms with the largest fee proposal calculation mistakes, rather than firms offering the best value. If fee must be the determining factor, then detailed, high-quality documentation must be provided for comparable bidding to occur.

One fee-based selection process uses an alternative method known as the "Middle-Bid" method, which can help reduce the chance of selecting an integrator that is either over- or under-priced.

A warning about Internet auctions

Some Internet sites have been set up to allow Internet-auction style bidding on AV integration projects, facility design and construction services.

AV design and integration is multi-faceted and highly technical. These services are very complex and require superior expertise.

Using Internet bidding as a procurement method in the design, construction and AV trades is a risky proposition.

While this option can work for purchasing equipment without design and installation, any design or integration services should be procured using the qualifications-based methods.

This process works as follows:
1. Create a detailed RFP, including a qualifications request or minimum qualification requirements for potential bidders.
2. If possible, invite a small select group of integrators (usually three or five) with the required qualifications and prior experience, rather than allowing a large number of bidders.
3. Bids are opened and the lowest and highest bids are set aside. The middle bid is then selected for the contract award.

This works best with an odd number of bid responses. This way there is only one "middle" bidder. With an even number of bidders (say, four or six), there are two middle bidders. In that case, there should be a predetermined selection criterion, e.g., the lower or higher of the two middle bids is the winner.

Structuring the RFQ and the RFP

The Request for Qualifications (RFQ) and the Request for Proposal (RFP) are basic structures for soliciting information and bids from potential service and equipment providers. Much of the information requested is similar to requests made to other typical building trades, but there are a few items specific to the AV trade that should be included in these document packages.
This chapter outlines the overall structure of the documents, while Chapter 5: The Design Phase offers more detail on the required content, particularly for the RFP.

Request for Qualifications (RFQ)

A first step in hiring an AV provider is preparation of the Request for Qualifications (RFQ). While RFQ's may be incorporated into a Request for Proposal (RFP), it is advisable to request qualifications first (or as a separate document if an RFP is also distributed). This allows for a preliminary review of qualifications of those who will be bidding — a critical component of a streamlined and successful selection process.

The components of an RFQ, listed here, are only slightly different for integrators and consultants.[14] Some requirements that may be requested from potential bidders are:

Project Size Requirements

Some larger projects may require experience with projects of similar size or type. An integrator whose largest AV project was a $150,000 installation may not be qualified to handle a $2M system installation. To avoid this type of mismatch, it is useful to limit bidders on larger projects to those who have experience on projects of similar size and scale. For example, bidders to be considered for an AV installation that is estimated to cost $2M could be drawn from integrators that have successfully completed at least one project of $1M AV system cost or higher in the past five years.

Company Size Requirements

Similarly, the current size of the company and its location can be a factor in evaluating qualifications for a project. Bidders may be limited to those with a minimum number of technical staff for instance.

Warranty Response Time

Once a project is installed, the owner may require a specified response time to have personnel on-site to address warranty issues as required. A maximum time (such as 24 hours or less, for example) may be required, as well as the specification that the bidder have a permanent office within a certain radius of the project site.

Additional information typically requested includes:

Corporate Profile
- Number and roles of full-time staff
- Corporate history
- Former company names
- Licensing
- Subcontractors — control system programmers and others
- Product representation, certifications and service authorizations

Team Personnel
- Key Personnel (primary and alternate)
- Project executive
- Project manager
- Systems designer
- Crew chief/superintendent/lead technician
- Systems programmer
- Commissioning agent
- Trainer

---

[14] See RFQ Checklist, end of chapter.

- Experience
- Length of employment
- Certification and licensing
- Qualifications
- Current workload

Project Experience Relative to:
- Scope and scale
- Dollar value
- Similar technology applications

Project References

Current Workload

Statement of Ability to Meet Schedule

Type and Extent of Facilities *(office space, warehouse space, staging area, service offices, etc.)*

Type and Extent of Electronic Test Equipment

Bonding Capacity

Previous Performance Bond Experience

It may be useful to know if the integrator has defaulted or otherwise failed to perform on a previous project. A requirement can be included to disclose if an owner was required to exercise a performance bond on a past project, when this occurred and why.

This form is used by the federal government on building projects to obtain standardized qualifications information from potential design team members. It is administered by the General Services Administration. To download and obtain more information on the SF330 (and its predecessors SF254 and SF 255), go to http://www.gsa.gov/ and search for "SF330".

Figure 14. Form SF330

ARCHITECT - ENGINEER QUALIFICATIONS

PART I - CONTRACT-SPECIFIC QUALIFICATIONS

A. CONTRACT INFORMATION

1. TITLE AND LOCATION *(City and State)*

2. PUBLIC NOTICE DATE | 3. SOLICITATION OR PROJECT NUMBER

B. ARCHITECT-ENGINEER POINT OF CONTACT

4. NAME AND TITLE

5. NAME OF FIRM

6. TELEPHONE NUMBER | 7. FAX NUMBER | 8. E-MAIL ADDRESS

C. PROPOSED TEAM
(Complete this section for the prime contractor and all key subcontractors.)

(Check) PRIME / J-V PARTNER / SUBCONTRACTOR | 9. FIRM NAME | 10. ADDRESS | 11. ROLE IN THIS CONTRACT

Financial, Legal and Insurance Information
- Trade and bank credit references (include Release for Confidential Matter)
- Dunn & Bradstreet rating
- Recent and current litigation experience, both project and non-project related
- Insurance limits

Scoring and Weighting Scheme
For requests that will be subject to a scoring process, it is important to include on what basis the scoring will be made, what the maximum score (or weighting) will be for each element evaluated, and how many of the top-scoring bidders will be placed on the "short list" for continued consideration. In some cases, there may be simply a minimum passing score that is required.

In order to standardize the responses, a form can be provided by the entity requesting the qualifications. Either a custom form may be used, or a standard one that is commonly used in the construction industry. While the standard forms are useful for the traditional trades, they may not specify all the types of information most relevant to gather from AV providers.

A common standard form is the SF330, which is often used on Federal projects and may be requested on non-Federal projects as well. The other common informational form is the AIA (American Institute of Architects) Form A305™-1986. See the AIA website (www.aia.org) for a listing of more AIA forms that apply to construction projects.

Request for Proposal (RFP)
The RFP package, a collection of documents used to request a response to the specific project, consists of:
- Details of the scope of work to be performed
- The administrative terms of the potential agreement
- The pricing for the scope of work by the AV provider

RFPs are generally prepared by the owner or architect to hire either a design consultant for a consultant-led option or an integrator for a design-build option.

In a design-bid-build process, the consultant normally prepares almost all of the RFP package for the integrator contracting process. (The details of this bid package, as well as a checklist can be found in Chapter 5: The Design Phase.)

The following describes the basic RFP components for either design or design-build services:

Project Description
A description of the project helps establish in a general sense what type of project it is and what the project encompasses, providing an overview of the scope of work. For an AV RFP, the AV provider should be given information about the architectural project size and budget, including any details that may have been developed for the architectural program, as well as any AV program information already existing. This information should be provided as a part of the RFP package.

Project Schedule

With information about the project schedule, the AV provider can evaluate the effort that may be required, depending on whether the schedule is tight or generous. There may also be special circumstances or events associated with the project that could have an impact on the design and installation process. Schedule information should include the overall architectural design and installation schedule, including end-user move-in and occupancy schedules when available.

Administrative Contract Requirements

Any terms and conditions that will be part of the contract if awarded should be included in the RFP. This may include standard language from industry-standard documents, or language developed by the owner for all projects. More information on some of these forms is provided in Chapter 6: The Construction Phase, as well as in the appendices.

General Conditions and Design Documents

If already developed, it would be important to include architectural plans and sections of either the entire project or at least the locations in which AV systems. For integration RFP's, there may be a set of administrative General Conditions as well as detailed system design information (drawings and specifications), if the RFP is issued under a design-bid-build method. More information on the structure and content of these documents is covered in Chapter 5: The Design Phase.

Using The Bid Process

The bid process usually applies only to design-build or installation-only contracts under design-bid-build. AV consulting services are sometimes selected under a bid process, but this is not recommended.

The bid process starts with the issuance of an RFP. Ideally, it is issued after a "short list" of potential bidders has been established under an RFQ. Otherwise, criteria should be included in the RFP documents that limit responders to those who are qualified, if possible.

Issuing the RFP

Generally, the RFP is officially issued by the entity that is hiring the service provider. This could be an architect issuing an RFP to consultants for design services or the owner issuing an RFP to integrators for design-build services. To select an integrator under an design-bid-build process, an architect often issues an RFP that has been prepared by an AV consultant to the general contractor; the GC in turn solicits an AV integrator using that RFP. This may sound complicated, but it can work quite well if the processes in this book are closely followed.

The details of the process are typically governed by the organization's own policies. The larger the organization, the less flexible it may be in handling an RFP. At the highest level, there are legal dictates, as in the case of some federal facilities where GSA and other department policies as well as legal requirements may apply.

While advertisements may be placed in appropriate publications to generate responses to an RFP, this is not the preferred approach. (On the other hand, requests for qualifications can benefit from the broad use of advertising channels.)

How the Bid Process Works

In the RFP, a date, time and place should be given for the delivery of the bid response. A reasonable amount of time should be allotted for bidders to issue a proper response. Bid preparation requires input from multiple sources, such as sub-contractors, manufacturers, and various departments within the bidder's organization. For a simple, straightforward project, two weeks should be sufficient. Larger or more complex projects deserve a response time of four or five weeks. Holiday periods should be allowed for when establishing response times.

Once the RFP is issued, the bidding process continues with a pre-bid meeting, issuance of written, bid clarifications, bid evaluation and project award. Throughout the process, consideration must be given to the timeline of the bid. The typical milestones for a full-blown bid process are:

1. **RFP Issued**

   The RFP may be distributed to a select group or announced to bidders, who then take action to obtain the RFP documents.

2. **Pre-Bid Meeting**

   Once the bidders have received and reviewed the bid documents, a pre-bid meeting is held. In addition to the consultant and bidders' representatives, the owner, architect, general contractor, and other key project participants should be present.

   The pre-bid meeting should include a walk-through inspection of the site, if it already exists. Alternatively, the architect, owner and/or consultant can use the design documents to provide a site review.

   During this meeting, bidders have an opportunity to examine site conditions, review physical logistics and obtain clarifications from the consultant/ design team.

3. **Cutoff of Pre-Bid Questions**
   During the pre-bid meeting (and, in fact, throughout the bid process), general questions or bid clarifications may surface. There may also be other changes to conditions or specifications that would affect the bid documents. In these cases, bid addenda should be compiled, including answers, drawings, and all other pertinent information. The bid addenda must be formulated and transmitted to all bidders in a timely manner.

   Unless the issuance of addenda warrants an extension to the bid response deadline, there should be a cutoff time for questions regarding the bid package. Generally, this is established a week before bids are due.

4. **Bids Due**
   In some government processes, the bids are opened and read aloud at the specified date and time. For most private organizations, the bid documents must be delivered by the specified time and will be opened later.

5. **Post-Bid Interviews**
   On some projects, interviews are conducted after the bids are received to further explore the qualifications and acceptability of the potential AV service provider.

6. **Decision for Contract Award**
   At the end of the bid process, a decision is made to award the contract to the selected bidder, and notification should be provided to bidders who were not selected.

Scoping Strategies for Contracting

There are important contract preparation issues under the different delivery methods that will arise at various points along the way. Here various approaches to contract preparation are examined. In addition to the contract variations with the two basic design-bid and design-build options, there is a potentially thorny issue of control system programming. See the appendix for a checklist of typical scope items for these types of contracts, as discussed below.

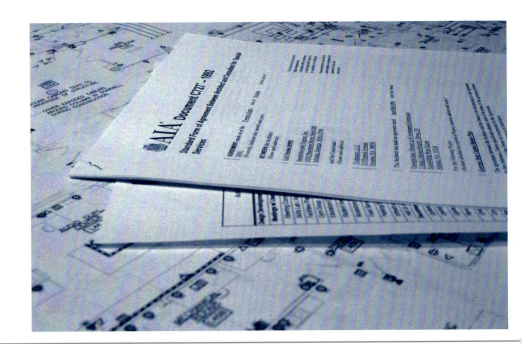

Design-Build Contracts

In this process, if the owner does not have an applicable contract template developed for AV services, the integrator may create the majority of the AV portion of the contract documents. By extension, in an integrator-led design-build project, the integrator is normally the only contracted AV provider. This is a fundamental difference between the integrator-led design-build process and the consultant-led design-bid-build process.[15]

Under the design-build method, the integrator is typically responsible for all phases of the AV work, including AV system needs assessment, design, installation, commissioning, training and service. The consultant-led design-build variation could include a separate contract with an independent consultant.

Sometimes, there is a separately contracted program phase prepared by an independent consultant under a consultant-led design-build process. Whether separately contracted or not, needs analysis is crucial for defining the functionality of the AV systems needed by the end-users. If the needs analysis has not been performed, the contract will be too vague.

While a full specification is not required as a basis for a design-build contract, a simple equipment quotation is insufficient. The integrator, or a consultant, should assess the end-users' needs, which can be the basis for the integrator's proposal, along with a detailed, easy to understand system description. Some technical details may be required, but a clear, descriptive narrative describing the finished system's functionality may be sufficient.

A scope of work, listing the responsibilities of all parties (owner, contractor, integrator and any other trades) should also be included. The following are the two primary options for defining a design-build contract scope:

Written Narrative/Non-Technical RFP

A non-technical RFP is a narrative that includes few specifics about the proposed equipment or its specific configuration. It is simply a narrative description of what the system needs to do. This document outlines the desired functionality of the AV system. It includes quality expectations, operational expectations, and a written description of what the systems are supposed to accomplish. Without the equipment, manufacturer, and installation detail, the document allows a great deal of flexibility in the means and methods the integrator can use to accomplish the system goals.

Although this may be required in a fast-track project, the brevity of the document allows too much latitude and contains insufficient contract-enforceable quality controls for the design and installation process. In the end, it would be difficult for an owner or end-user to answer the question, "Am I getting what I paid for?" While this type of RFP can be used effectively under a consultant-led process as a component of a full bid package, a technical RFP is always a better choice for design-build.

---

[15] InfoComm has developed a standard contract template for contracting design-build services between an integrator and an owner or architect, available from its website at www.infocomm.org.

Technical RFP
A technical RFP should include a narrative of the system functionality and broad schematic diagrams, showing function of sub-systems and/or major components. With the exception of major components, the systems integrator still has considerable latitude in the methods used to design and install the system.

A technical RFP should also include written specifications that define the overall system quality expectations and individual components, construction procedures, submittal process, substitution process, RFI process, acceptance testing and commissioning.

Design-Bid-Build Contracts
The design-bid-build process can be described as "establishing one contract to write another." An AV designer, usually an independent consultant, is contracted to create the bulk of the documentation whose final result is a contract between the installing integrator and the owner or building contractor. The two contracts involved are:

1. The Consulting Contract
In a consultant-led process, the consultant provides the pre-installation design consulting usually including a program phase. In addition, the consultant provides services during the construction stage in much the same way an architect or electrical engineer does in building design projects. Construction administration, including site visits and submittal reviews, should be included in the contract. In addition, system commissioning and end-user training, critical to the AV implementation process, should be included.

2. The Equipment Procurement and Installation-Only Contract
For the integrator's installation-only contract that follows the design contract, the consultant prepares most, or all, of the contract documents. This includes the design package as described in Chapter 5: The Design Phase. Under this option, drawings and specifications are included in the contract documents that clearly define the system's configuration and equipment, as well as procedures, submittals, testing and quality control measures — enforceable within the contract limits. This option restricts the integrator's latitude in the configuration and installation of the system. It shifts more design responsibility to the consultant while adding necessary process checks and balances.

Program-Only Contract
If a program report is unavailable or inadequate, a contract to fully develop the report is required before the scope of a design or installation contract can be determined. This process would most often be required under the following circumstances:

- No program report has been developed and the project is a mid-size to large project (greater than approximately $150,000).
- The needs are vague or poorly defined.
- The budget available for the known requirements is inadequate.
- A program document exists, but it needs to be verified, either because it is outdated or contains insufficient detail for an AV scope of work to be adequately defined.

When hiring a provider to conduct only the needs analysis, the best practice is to hire that same AV service provider for the entire project — assuming it is relevant to his or her scope or competency, i.e., a consultant would be hired for design, construction monitoring and commissioning; an integrator would be hired for design and installation or just installation, depending on the method used.

The resulting analysis should provide sufficient information for a scope of work to be defined, a budget for the systems to be established, and a proposal to design and/or install the systems. See Chapter 4: The Program Phase for details on creating a program report.

An integrator may conduct a programming process without charge if the project is not very large and the information for what is needed is readily available. This analysis may be incorporated into a proposal for design-build services. If, however, the project is very large and/or complex, most integrators will charge for this service — sometimes crediting it toward a design-build contract if they are hired for the remainder of the job.

The program contract can be part of any of the consultant and/or integrator based delivery methods. (All independent AV consultants provide needs analysis as a part of their standard offerings since it is an integral part of the design process.)

The result of this contract is information that is normally used to proceed into the design phase of a consultant-led design-bid, or a consultant- or integrator-led design-build process.

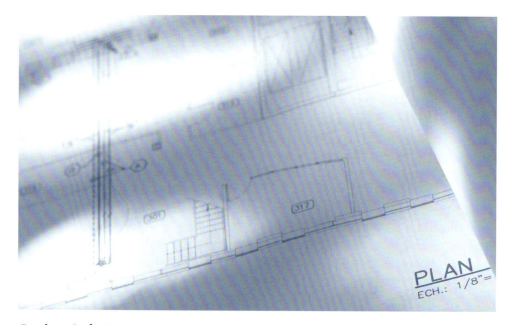

Design-Only Contract

In a consultant-led process, the initial contract to be established is the contract for the program and/or design phase that should also include construction administration and monitoring, as well as commissioning and training tasks. With the exception of system training, the range of services in the AV design-only contract is similar to the typical A/E contract for building design, in that it covers consulting services for both the design and construction phases of the project.

Sometimes, however, even the architect is not contracted for the construction stage. The design is contracted without the follow-on services normally provided by the design team. When this happens, it may be because the owner needs to raise funds for a project before proceeding into construction or is uncertain it will be built at all.

There are also occasions where an integrator is requested to enter into a design-only contract. Under a design-build option, this is usually done with the idea that the same integrator will be providing the installation for the system.

BEST PRACTICE

Stick with the same provider after a design-only contract

Under a design-bid-build option, it is best to contract initially with an independent consultant and include construction administration services in the contract to begin with. If the construction phase services are not included, then the same consultant should be hired for the remaining services if the project continues. Consistent technical construction administration can be a determining factor for success of the AV system.

Under an integrator-led alternative, following through with the same integrator for the systems installation is critical. When contracting with an integrator during the design phase with a design-build path in mind, don't "shop" the design documents to other integrators at the end of the design phase in the hopes of reducing cost. The documents are not likely to be appropriate for comparable bidding and may lead to inconsistent or incorrect pricing; they also may not convey the owner's full expectations for the systems performance.

In either case, the discontinuity that results from "changing horses in mid-stream" will undermine the advantages of the delivery method being used.

The result of a design-only contract should be either a complete set of bid documents to be used in a bid process or, perhaps, a negotiated process with a pre-selected integrator or an add-on contract for a designing integrator to install the systems.

A particularly difficult situation is created when one integrator is hired to design an AV system while another is awarded the project. The potential problems are:

- The integrator awarded the contract may be a competitor of the one who "lost" the project. Consequently, the two firms may not communicate well if questions arise.
- Specifications may not have been developed if the designing integrator was unaware that the installation would be offered to other integrators. This creates unenforceable quality control in the installation contract.
- Since the designing integrator is usually not contracted to provide construction monitoring and oversight, the new integrator may have many unanswered questions and the owner's staff may not have the time or experience to assist during construction.
- If the designing integrator is contracted to provide construction administration services, then conflicts of interest may arise, or information sharing and oversight may be difficult — if not impossible.

AV Integration Contract Issues

The structure and terms of the contract can have a significant impact on the AV integration process and the quality of the installed systems. Below are specific issues that should be considered in structuring AV integrator design-build or installation contracts.

Avoid the "Double-Sub" Contract for AV Integration

Before the advent of modern low-voltage systems, electrical contractors handled all electrical work on a building project — both basic power and specialty systems. Today, specialty systems required in modern technology-rich buildings are often highly sophisticated, and their range, volume, and complexity demand more focused expertise. Numerous specialty contractors now provide installation for a host of electronic systems, including pro-AV.

Some specialties, such as fire alarm and life safety, fit comfortably inside the traditional subcontract models and project schedules. However, the size, scope and schedules required for professional AV systems make these types of installations less suitable for sub-subcontracting on a building installation team.

Thus, it may be best (particularly on large AV projects) to subcontract the AV integrator to the general contractor or contract directly to the owner for installation.

No matter what the situation, subcontracting the AV integration under another subcontractor such as electrical or data/telecom is not recommended, for the following reasons:

1. Though cabling can be pulled before construction of the site is completed, the AV integrator cannot begin the bulk of the AV work until the construction is complete, and the site is clean and dust-free. The racks of electronics, projection screens, cameras, microphones, and control systems cannot be installed until the facility is clean enough to avoid damage. This installation occurs at the end of the construction phase when the general contractors and base building sub-contractors are done with all the heavy work and are primarily addressing punch list items.
2. On large projects, the AV contractor may spend weeks completing the installation and providing training for the new systems' users. This extended timeframe makes sub-subcontracts unwieldy, since they must stay open even though the prime contractor is trying to close out the project.
3. Sub-subcontracts can establish an unnecessarily extended chain of command between the integrator and the owner and end-users. AV installation can be made more difficult if there is a lack of direct contact with the owner's staff to coordinate room schedules, local area network operation and communications services.
4. There are cost markups associated with sub- or sub-subcontracts in building projects. With markups between five and ten percent, the costs can be significant on large AV systems, especially those in excess of a million dollars. Since the bulk of AV installation work occurs after the traditional contractors are mostly done at the jobsite, the coordination required by a GC or another sub-contractor is much less than required for the traditional MEP contractors who are almost constantly onsite during the entire construction phase. Thus, the usual five-ten percent markup is less justified than it would be for the other trades, and an owner might save money and increase efficiency by contracting with the AV provider directly.

The following table summarizes the pros and cons of contracting the AV integrator to the electrical subcontractor (the "double sub" condition), the general contractor and directly to the owner.

BEST PRACTICE

Contract AV integration directly when possible

When all is said and done, a contract directly with the owner is the better option for AV integration. Since the AV contractor may not be engaged until the general contractor's work is well underway, a coordination allowance may need to be established in the general contractor's contract before the AV integrator is brought on board. However, this should be substantially less than the traditional contractor markup for a standard subcontract arrangement, especially for larger AV projects.

AV Integrator Contracted to:	Pros	Cons
Electrical Subcontractor	▪ Single point of contract responsibility through GC	▪ Double contractor markup to owner ▪ More layers of responsibility and communication ▪ Contract closeout may be complicated by extended AV integrator time on-site after base building substantial completion
General Contractor	▪ Single point of contract responsibility	▪ Added layer of contractor markup ▪ Contract closeout may be complicated by extended AV integrator time on-site after base building substantial completion
Directly to Owner	▪ Relationship at end of project easier to manage ▪ Saves on GC and other contractor markups	▪ Multiple points of contract responsibility ▪ More owner coordination required

Chapter 3 — Selecting and Contracting the Project Team

Avoid Separation of AV Cabling, Equipment and Systems Contracts

Due to the staging of large construction projects, the AV cabling is sometimes installed when the ceiling is accessible and is treated as a separate contract before the final system design is complete. This may occur in buildings in which room-to-room cabling is required — either from system to system or between systems and a control or interconnection room (sometimes called a master control room).

In this case, a separate cabling contract can be prepared with the following considerations:

Include Only Room-to-Room Cabling

For a separate cabling contract, include only the cabling that will be difficult to install later in the project. Generally, this means cabling that will run down corridors, perhaps to a riser or control room. The systems design, however, must be far enough along for this cabling to be known. Cabling that is required within a room system, such as from a lectern to a projector or from wall boxes to a local equipment rack should be deferred to the system contract.

Use the Same Integrator for AV-Specific Cabling and Systems

For the separate contract, the best practice is to use an AV integrator for the AV cabling, particularly for cabling such as RGBHV or SDI video that is not normally part of a structured cabling data contract. In this case, the integrator who will be doing the system installation should install the cabling. That way, the same integrator will be dealing with the cable once the systems are connected to the cabling network.

The Use of Structured Cabling for Transporting AV Signals

Structured cabling systems[16] include the use of standardized twisted pair data cabling and optical fiber for computer and telecommunications cabling. Audio, video, and control signals can also make use of this cabling either by installing proprietary cable interfaces or by using a computer data network to transport the signals. This may be an advantage on some projects.

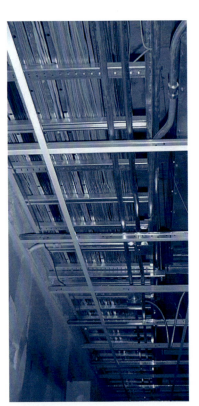

Contracting AV-Related Structured Cabling

In some cases, the final systems may not be completely developed, but room-to-room cabling needs to be installed early in the project, so some structured cabling is included in the data/telecom design to accommodate the anticipated AV pathways. In other cases, there may be plans from the start to use structured cabling to provide connectivity between systems. For these instances, it is often advantageous to allow a project data/telecom contractor to install and test these cables.

Complete testing and signoff of the installed cable should be required — just as it would be for the rest of the structured cabling system to facilitate the hand-off to the AV integrator when termination of the cabling is required.

Avoid Splitting Equipment and Labor Contracts

If a project involves purchase of equipment that is to be installed by the owner's organization, then line item bids for the equipment alone are acceptable. However, problems may occur when new AV equipment is purchased under one contract and installed under another, especially if the purchases are from two separate firms (which has become more likely with the availability of pro-AV purchases through the Internet).

---

[16] Refer to BICSI for information on structured cabling systems, www.bicsi.org.

If the contract is split between two firms, warranty issues and the difficulty in determining responsibility for system problems can create undesirable confusion. If equipment is to be purchased separately from the installation services, as may be desirable for certain tax-exempt owners, it is best to purchase both equipment and labor from the same integrator. The fact that there is a single point of responsibility for the two contracts mitigates potential problems.

Contracting for AV Software

An increasingly important issue in contracting pro-AV systems is software licensing and ownership. With AV systems utilizing Digital Signal Processing (DSP) and LAN-based configurations, and with control systems incorporating graphical user interface (GUI) options, software has become as important to successful AV system integration as hardware.
The AV industry, the building industry, and facility owners are grappling with finding the best ways to address developing and contracting for the software components of AV systems. This problem is often encountered when an owner wants to use software (particularly control system software) that was developed and installed for one project for another similar project that may have been installed by a different integrator. Problems can also occur for consultants who design a user interface that the owner wants to use on subsequent projects that may be designed and installed by a different AV team.

Owners who want flexibility in upgrading current and future facilities want to own the code so that they can reuse it and provide it to other integrators. Integrators, on the other hand, want to own the software they develop and license it to the owner for that specific project.

There is validity to both sides of the argument. Owners who have multiple locations (e.g., college campuses) need to be able to reuse a standard graphical user interface design from one room to another or from building to building. Limited by a software license, they may not be allowed to do so without using the original consultant or integrator. While this may be acceptable in many circumstances, it could be difficult if the original providers did not perform well enough to be considered for the new project, or if the new project is outside the scope or geographical area of that provider.

On the other hand, consultants and integrators deserve remuneration for the development and distribution of their original project software creations.

The most common agreement is that the software for the project at hand cannot be reused or modified for any other project. This mirrors the licensing agreements for off-the-shelf computer software, such as word processing and spreadsheet packages. To ease potential conflicts, all parties need to develop contract language that is appropriate to the project and the client involved. The possibilities range from:

- Providing complete ownership and usage by the client of all software developed
- Licensing the software and preventing the use or modification of control system software, user interfaces or other custom AV software
- "In-between" option of licensing the user interface look-and-feel for the project with a fee to be paid when it is reused on other projects, but with more restrictions on the operating code behind it

Ongoing efforts in the AV industry are aiming to address these issues in the form of standardized contract and licensing agreements, some of which are referenced in later chapters of this book.

Contracting Control System Software Programming

Even though the AV designer lays out the control system touch screen or web interface design, a programmer will be required to implement the operating code that responds to the user interface and makes the system work.

Because of the importance of the control system functionality to a successful AV system, how the control system is programmed is a key factor. The integrator is usually, but not always, responsible for the control system programming by contract to the owner or the GC. However, a number of contracting options are available:

In-House Integrator Programmer

In this traditional method of completing control system programming, the integrator has a staff of in-house programmers who handle most if not all of its projects. This option gives the integrator more control over the programming process and the programmer.

Because this is a relatively new specialty, programmers may vary in their programming expertise or their knowledge of various vendors' software languages. In addition, the programming may not be the only job function of the programmer. This means less time is applied to the project. Having an in-house programmer, if those resources are reliable, does mean that warranty or service issues can be addressed more promptly than if the programming is outsourced.

Independent Programmer Subcontracting to Integrator

The next most common arrangement is for an integrator to subcontract to an independent control system programmer for the operating software and user interface design, often under a design-build process. This option allows the integrator to obtain specialized expertise. More independent programmers are available and more control system vendors are offering certification programs. This helps to provide a basic metric for the individual programmer's level of skill and expertise. Although this is becoming an attractive option in many projects, there is a potential licensing issue with more sub-contractors and more contracts.

Independent Design Consultant Subcontracting to Integrator or Owner

As independent consultants have become more involved in programming various parts of the AV system due simply to technology trends, they are beginning to offer control system programming services in addition to the traditional facility and system design services. This option has an advantage in that the programmer has familiarity with the particular project and has an understanding of the end-user's needs.

The potential problems with this situation depend upon the contractual arrangement. There is a potential conflict of interest if the design consultant and integrator are both contracted to the owner and the same consultant is contracting to the integrator for the control system programming. The design consultant is still representing the owner's interest in the contractor's performance, but is put in a position of being paid by the contractor for the programming work. Therefore, this arrangement is not generally recommended.

A better arrangement is one in which the independent design consultant contracts directly to the owner for control system programming. This situation is similar to the independent programmer option.

Independent Programmer Direct to Owner

Another viable option is for an independent programmer to be contracted directly to the owner when the integrator is not providing programming. This can be useful for simplifying (but not eliminating) the control software licensing issues and providing more flexibility in the owner's relationship with either the integrator or the programmer. It does, however, mean that the owner must provide more contract and project coordination, especially under design-bid-build scenarios where there could be three AV contracts to administer: the consultant, the integrator and the programmer.

Owner as Programmer

In more advanced AV organizations with a large number of AV systems to manage, a part-time or full-time control system programmer may be a member of the owner's staff. In this case, the programmer works with the integrator to program the hardware. Albeit a potentially more difficult working relationship, it can eliminate licensing or ownership issues. The integrator may be required to license the owner's code for support purposes. The disadvantage of this option is that AV system warranty and service issues can be unclear in determining whether an AV problem is a software or a hardware issue.

There is no clear best practice among these options. Each situation should be carefully evaluated with regard to the project size and type, the delivery method, and the particular owner, consultant, and/or integrator involved.

The first two options are the most common and have the most history in the industry. The others are becoming more viable and the best practices for these will continue to evolve as certification and licensing issues are addressed.

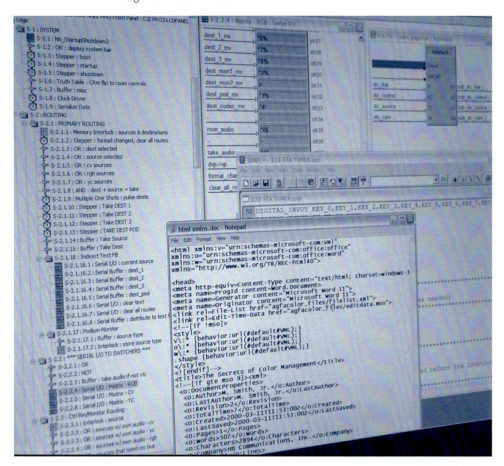

Chapter 3 Checklists

AV PROVIDER SCOPE OF WORK CHECKLIST

This checklist provides some typical scope of work items that may be included on many AV design and installation projects. This list includes items in either a consultant contract under a design-bid-build process or a design-build contract with an integrator. A few items only apply to one process.

Items with no highlighting apply could apply to either an AV Consultant or an AV integrator under either process.

Items that apply only to an AV Consultant under a design-bid-build process are highlighted in dark blue.

Items that apply only to an AV Integrator under a design-build process are highlighted in light blue.

Note that this checklist is just that — a reminder to address items that may apply to a particular project. All items may not apply, and there may be other items that are desirable or required for a particular project. More text should be provided for the items to fully describe their meaning in a proposal or contract. This list is not intended to be used as a contract template, but is merely a guide to its potential structure. Any proposal or contract should be reviewed by a qualified attorney before signing.

Provide a project summary, including information about the overall project and specific spaces to be addressed for AV, if needed for clarification.

AV System Programming Services
- Prepare preliminary program questions
- Conduct an AV Program Meeting
- Prepare an initial AV Program Report
- Prepare preliminary opinion of probable costs for the AV systems
- Conduct an AV Program Review Meeting
- Modify the Program Document if required
- Attend (X) additional meetings

Schematic Design Phase
- Attend (X) coordination meetings
- Provide space planning assistance pertaining to special AV spaces to address room orientations, furniture layouts, image sizes and sightlines
- Provide preliminary input to the design team concerning AV-related acoustical and lighting criteria and design
- Review and comment on schematic phase drawings for AV related infrastructure issues
- Assist with preliminary base building infrastructure budgets

Design Development (DD) Phase
- Attend (X) coordination meetings
- Refine preliminary equipment layout sketches to address sightlines, projection throw distances, equipment locations and general presentation orientations.
- Provide additional input to the design team concerning AV-related infrastructure including:
  - Cooling requirements for AV equipment
  - Power and grounding requirements
  - Data/telecom requirements
  - AV-related millwork
  - Room acoustics and noise control designs
  - Lighting designs
- Provide AV signal conduit and backbox layouts for AV systems
- Perform a general review of the design team drawings and specifications to address coordination issues related to AV infrastructure requirements.
- Assist with preliminary base building infrastructure budgets

Base Building Construction Documents (CD) Phase
- Attend (X) coordination meetings
- Provide final input and coordination to the design team concerning AV-related infrastructure including:
  - AV-related millwork
  - Power and grounding for AV equipment
  - Data/telecom requirements
  - Room acoustics and noise control designs
  - Lighting designs
- Provide final AV signal conduit and backbox layouts for AV systems
- Perform a general review of the design team drawings and specifications to address coordination issues related to AV infrastructure requirements.

AV System Construction Documents Phase
(Note any scheduling issues such as the issuance of the AV system design package after base building construction documents are complete, if applicable. These scope items may fall under the Base Building CD Phase on some projects)
- Provide ongoing coordination and review with the design team
- Prepare AV System designs in accordance with the approved AV Program Report and provide to owner team for review

- Prepare a final detailed opinion of probable cost including an AV equipment list
- Attend (X) coordination meetings
- Provide (X) copies of final specifications and drawings for the AV systems for solicitation of bids from qualified AV system contractors.
- Provide (X) copies of final AV system design drawings for the AV systems to the owner before installation

AV System Bidding Phase Services
- Establish a list of recommended pre-qualified bidders
- Evaluate potential bidders suggested by the Owner
- Coordinate the issuance specifications and drawings to bidders with the project team
- Conduct a pre-bid meeting.
- Conduct a site tour (if applicable)
- Review and comment on questions and requests for information during bidding
- Prepare input for document addenda, if required.
- Review bids and make recommendations for award
- Conduct integrator inteviews

Construction Phase
- Attend (X) construction meetings
- Conduct one meeting with the electrical contractor to review the AV conduit and power requirements for the project prior to installation.
- Provide on-going coordination regarding AV related base building elements.
- Make up to (X) site visits at key milestones in the construction process to monitor AV infrastructure installation
- Review and comment on requests for information.
- Review and comment on change order requests.
- Review AV integrator submittals
- Provide detailed AV installation schedule to installation team
- Provide (X) copies of final as-built documentation at completion of AV system installation

AV System Commissioning & Training
- Review the AV integrator's system test reports.
- Perform a preliminary checkout of the prototypical AV systems, if applicable.
- Perform a final checkout of all AV systems and prepare punch lists.
- Conduct tests and provide final alignment of the AV Systems
- Conduct (X) training sessions with the technicians and end-users
- Provide recommendation of final acceptance of the AV systems at completion

Warranty Phase
- Provide warranty service for (X) months after substantial completion
- Provide (X) preventive maintenance visits to the site

Potential Additional Services
- Attend additional meetings
- Revise the approved AV program report and/or system budgets based on changes requested by the Owner after program approval.
- Revise AV conduit after 100% Base Building Construction Documents, based on Owner or Client-requested design changes.
- Provide design services for additional AV systems
- Provide installation services for additional AV systems
- Provide extended service period on equipment

RFQ Checklist

This list represents items that should be included in an RFQ. More or fewer items may be required, depending on the project and the context in which the RFQ is issued.

Project Information and Response

Requirements
- Project Description and Schedule
- General Scope of Services Anticipated
- Response Due Date, Time and Address
- Any minimum size of project that must have been completed within a recent time period
- Any minimum company size required to respond
- Warranty response time required
- Maximum allowed distance from project site to responder's nearest office
- Criteria and weighting scheme to be used for scoring responses
- Selection schedule including short list, interview and final selection time periods

Information being requested:

General Company Profile
- Contact Information
- Number and type of full-time staff
- Corporate History
- Former Company Names
- Office performing work if not main office
- Licensing
- Product Representation, Certifications and Service Authorizations
- Facilities
- Test Equipment

Financial and Legal Information
- Balance Sheet
- Income Statement
- Average Gross Receipts Past 5 Years
- Trade and Bank Credit References with Release Form
- Dunn & Bradstreet, if available
- Bonding Capacity
- Performance Bond ever exercised?
- Projects ever not completed?
- Insurance Limits
- Recent and current litigation experience, both project and non-project related

Project-specific information

Project Experience
- Scope and Scale
- Dollar Value
- Similar Technology Applications
- Similar Vertical Markets
- Subcontractors used for installation, programming or other work on this project
- Current Company Workload
- Ability to Meet Schedule
- Project References

Proposed Project Team

Key Personnel (Primary and Alternate)
- Project Executive
- Project Manager
- Systems Designer
- Crew Chief/Superintendent/Lead Technician
- Systems Programmer
- Commissioning Agent
- Trainer
- Experience
- Length of Employment
- Certification and Licensing
- Other Individual Qualifications
- Current Individual Workload

- Other company information at the discretion of the responder, if desired

CHAPTER 4
The Program Phase

Step One: Review the Existing Documentation and Facilities
Step Two: Benchmark Comparable Facilities
Step Three: Conduct the Program Meetings
Step Four: Write the Program Report
Step Five: Distribute the Report
Step Six: Approve the Program Report as a Basis for Systems Design

CHAPTER FOUR

With the vital information on project roles, processes, and contracts already covered, discussion on the AV integration process can begin in earnest.

AV projects begin with a vision. Individuals want or need to communicate more effectively with others and determine that AV technology is the vehicle to do so. A corporate executive needs to communicate company performance at a board meeting. A sales department needs to routinely communicate goals to a distribution network. University lecturers need to more fully engage today's students.

The vision of powerful communications will reach its full potential through the application of the processes described in this book. As explored in this chapter, the process of creating the AV reality begins with an AV professional's assessment of the needs and desires of the end-user — a needs analysis or program.

The necessary steps are laid out sequentially based on how the program phase should be conducted. It covers everything from the initial meetings and benchmarking to budgeting and reporting.

WHAT IS NEEDS ANALYSIS?

Needs analysis is the most critical stage of the design process because the results determine the nature of the systems, their infrastructure, and the system budget[17] including the impact of the expense on the base building.

The goal is to define the functional requirements of the AV systems based on the user's needs, desires and applications. Merely developing an equipment list is not enough. While the equipment list is an essential part of the design and installation process, it is not part of the needs analysis process. It is tempting, particularly from a box sales perspective, to go straight to an equipment list when determining the user's "needs," but only by conducting a needs analysis will one make sure that the user's needs are fully addressed by the final equipment list and system design. The AV team must capture and fully understand these needs before the system is designed.

Formal needs analysis consists of identifying what activities the end-users need to perform, then developing the functional descriptions of the systems that support those needs. During the needs analysis program phase in the building, design and construction industries (which is known as programming in the architectural trades), the AV professional determines the end-user's needs by examining the following:

1. The required application(s) based on the user's needs
2. The tasks and functions that support the application
3. The wishes and desires of the end-user

The result is a document, called the *program report*, that delineates the overall functional needs of the AV system (including budgets) based on the tasks the owner requires for operations.

[17] Large AV systems are often placed as budget line items (the cost of the electronics and related installation) but have a direct impact on the other trade budgets, e.g. architectural changes to accommodate additional space and/or greater floor-to-floor structural height than originally considered.

NEEDS DETERMINE APPLICATIONS

Applications support end-users' activities and needs. For example, the applications of training, teaching, and videoconferencing support a sales department's need to keep its sales staff informed and on target.

AV technologies often enhance existing end-user applications. These include classroom teaching, adult continuing education, training, corporate presentations, and data delivery for trading floors. The tasks involved in these applications are supported by the room and systems functions and features.

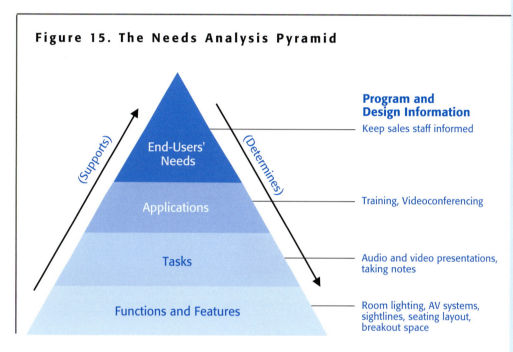

Figure 15. The Needs Analysis Pyramid

The Needs Analysis Pyramid illustrates both the top-down and the bottom-up concepts of examining the end-users' needs. Their needs determine the applications that support the need, which, in turn, determine the tasks and the functions required in the AV systems. The functions and features of the space (architectural, electrical, and mechanical), as well as the AV systems are then determined from this information. The labels to the right of the triangle give examples of AV-related data that is associated with each level.

FORM FOLLOWS FUNCTION

The basic premise for all building design is that form follows function. Determine the functions for a given building, as reported by the future users, and the building's form should be designed to accommodate these functions.

The same is true for pro-AV systems. Once the functions have been defined, the system with its infrastructure can be designed.

Architectural vs. software programming

For architects, "programming" is the process by which the overall requirements for the building are defined. Architects document the needs of the owner and end-user as a preliminary step to putting lines on paper that represent the space plan. The architectural program document discusses the facility in terms of square footage, space configuration and the overall quality of the building.

For AV professionals, "programming" often refers to software coding of programmable AV devices, such as control systems and digital signal processing (DSP) equipment.

Figure 16. Translating the Needs into a Design

An overview of how the end-user's needs ultimately define the design team's ability to create the required AV system electronics and how these have an impact on the architectural, electrical, and mechanical infrastructure design.

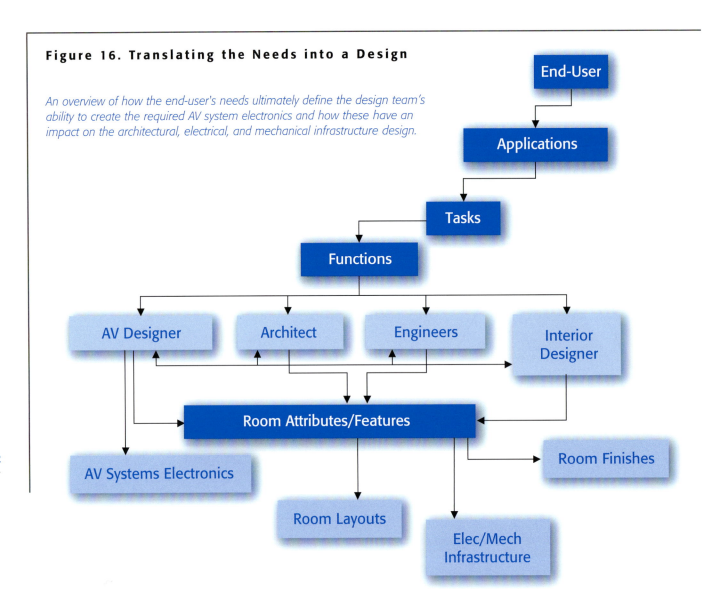

INCLUDING THE END-USERS

It is critical to include the end-users in the program process. In this case end-users are the people who will be using the system, as well as their managers and administrators. The owner and the end-users are needed to provide input to the design team to ensure their needs are met and that they are "invested" in the final product. Systems that are designed solely on design team assumptions are frequently off the mark.

To be effective, the process must take into consideration the requirements, the "desired applications," and the end-users' expectations — in short, the end-user's wish list. There should be a clear understanding and agreement on what the owner/end-users want to accomplish by the end of the design and installation process. Prioritizing at this stage helps to determine what actually will be accomplished, especially if the budget is insufficient to do it all.

The end-user may also be considering future requirements which will have an impact on the functions, features, and tasks that support current applications. An example might be a university with a popular on-campus course that it would like to offer off-campus via videoconferencing. While it may not be a current need, it may be included as an application in forming the AV systems program — budget permitting.

AN OVERVIEW OF THE PROGRAM PHASE

The first step in the process includes reviewing the available documentation and identifying the end-users who will participate. Meetings with the end-users are then set up, and the fact-finding begins.

The information gathered is interpreted and presented in a program report. Once this document is distributed, reviewed, and approved, it becomes the basis (sometimes part of the contractual basis) for the subsequent design phase. This review and approval process may take several rounds to complete to obtain owner and/or end-user sign-off as shown in Figure 17 below.

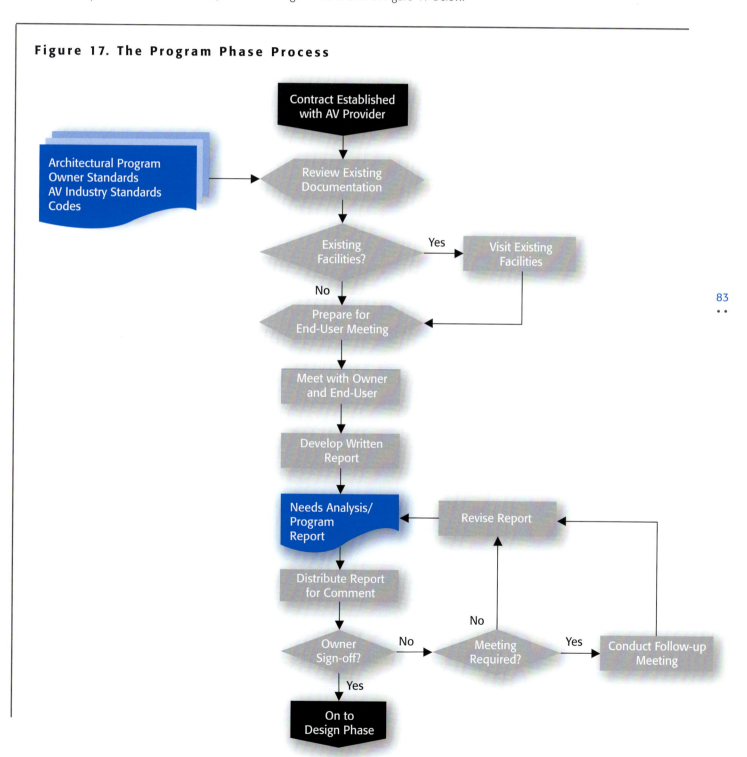

Figure 17. The Program Phase Process

CONTRACTING THE PROGRAM PROCESS

The contract for the AV program phase can take several forms depending on which delivery method is being used. Except in the case of an owner conducting an internal programming effort, a contract is usually required to start the process (see Chapter 3).

Typically, if an integrator- or consultant-led process is being used, a contract is signed to develop the program report. The contract might cover preparation of a new program or verification of an existing program. If programming is contracted separately from other phases, a follow-on proposal would normally be developed for the design and installation of the systems upon conclusion of the program phase. If, on the other hand, sufficient information is available, development of the AV program may be included in a broader contract for design or design-build services as required for the project.

If all or part of a program document is already available but needs to be updated or augmented, verification of the document may be required. For example, program verification will need to be performed if the owner or architect has developed some information for the AV systems but for some reason it is inaccurate, incomplete, outdated or not in a form that can be used as a basis for design.

No external contract is required if the owner is undertaking an internal program process. In this case, the group or individuals to be interviewed work directly with the internal AV professionals conducting the analysis.

Step One: Review the Existing Documentation and Facilities

Documentation is generally available with information about the existing physical, organizational and technical aspects of the project. If the AV systems are to be installed in an existing facility, it is important to tour these areas during the programming process to gather information about the physical aspects of the spaces and how they are currently being used.

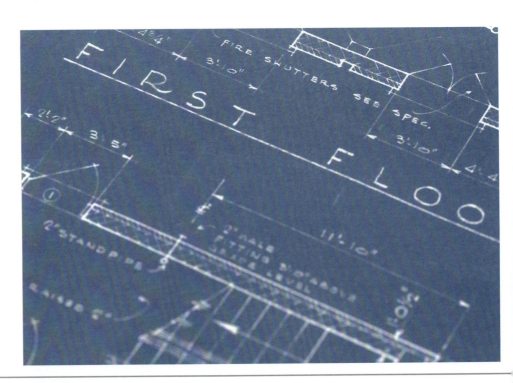

Existing documentation may include the following under each category of information:

Architectural
Architectural Program Document
The architectural program contains information such as the number of proposed classrooms, conference rooms or auditoria, the sizes of the spaces, and the number of people they will support. This can include what the architectural team has considered regarding AV systems in terms of space allocation and budgeting. The program also offers insight about the owner's operational needs and how those needs translate into space requirements.

Architectural Drawings
These drawings may be "as-builts" for an existing building or drawings of a design in progress. They may be available in paper and/or electronic form. The drawings depict the size of the spaces, their relationships to one other, and planned finishes. Floor plans, reflected ceiling plans, and sections of the spaces or buildings provide a good idea of the working spaces.

Engineering Drawings
These drawings depict the mechanical, electrical, structural, plumbing, and other engineered systems in the building. They help to identify areas of concern in terms of power, conduit, and data infrastructure, as well as any obvious HVAC noise concerns.

Organizational
Project Directory
This is a standard document that is produced and maintained by the architect, the owner or a construction/program manager. It lists the major contracted entities on the design and construction team, their identified role, and the primary contact.

Contract Scopes and Roles
The architect's contract delineates his or her role and may include terms that apply to the AV provider's role and responsibilities.

Owner and End-User Information
Background and operational information crucial to the AV project's purpose are often contained in the end-user organization's resources, such as websites, procedural and operations manuals, organizational charts or university catalogs.

Technical
Owner Standards and Design Manuals
These are documents that are not specific to a project, but include standards, criteria and procedures that must be adhered to in any project involving the owner's organization.

Codes and Ordinances
The most common codes that are of interest here are electrical, fire and life safety. In addition, there is a need to comply with the Americans with Disabilities Act (ADA) requirements. Many federal regulations have international equivalents and local authorities may impose additional restrictions.

Industry Standards

There are basic standards and recommended practices regarding audio, video and control systems. Additional research for specialty systems or certain vertical markets may need to be performed. For instance, interfacing with an audio and video broadcast system or healthcare equipment may require adherence to audio, video, and digital standards that are not typical pro-AV system standards. Other non-electronic requirements, such as the ANSI S12.60 Classroom Acoustics standard, may be required by some owners within the education arena.

Step Two: Benchmark Comparable Facilities

Visiting other similar facilities for review and comparison is often called benchmarking. This activity helps give the owner and the design team a common (and sometimes expanded) vision of what the user wants and needs.

Seeing a number of locations of similar size, type, and usage establishes a benchmark or guide on which to base the new facility design. Benchmarking offers the following benefits:

- It provides an opportunity to see varying approaches to design vs. budget.
- It may inspire new design ideas.
- The team can identify successful (and unsuccessful) designs and installations with regard to the project at hand.
- It can help to determine which functions and designs are most applicable to the current project.
- It allows project stakeholders to establish a communication path with other building managers and end-users about what they learned in going through the design and construction process and to discuss what they would do the same or different if they had to do it over again.

Benchmarking involves the following steps:

1. Determine appropriate facility types to visit. These may be a precise match to the owner's operation or they may be facilities with similar functions and similar operational needs.
2. Create a list of potential facilities to visit.
3. Check on whether or not the benchmark sites allow visits of this type. Some benchmark visits may only require a "user's" perspective in a public facility, but many are private facilities that require permission to enter. In addition, most benchmarking tours benefit from a behind-the-scenes tour that may require coordination with technical staff.
4. Narrow down options to a final list.
5. Determine who will go. The benchmark group may draw from the end-users, the owner's technical staff, the owner's administrative managers and the architectural design team, as well as the AV provider.
6. Schedule and make the visits.
7. Write a benchmarking report summarizing the sites visited, the pros and cons of each site, what impact there is on the client's anticipated needs, and the resulting AV systems that will support those needs.

Step Three: Conduct the Program Meetings

Program meetings should include representatives of both the design team and the owner. The purpose is to gather and exchange information in real time that will help determine what functions are required to support the end-users' applications. This information should reveal what they currently do, what they need to do, and what they want to do. To be effective, it is important to include all the key players in the meetings:

Attendees Representing Owner and End-Users

- End-Users — including representatives from each organization or department that will use the new systems. If the facility will be used by people outside the owner/operator's organization (such as videoconferencing facilities for hire or a classroom building serving several departments), the perspectives of those groups should be included whenever possible. If there is an owner or end-user technical committee (see Chapter 1), it should also be represented in the meetings.
- AV Technology Manager — representing the technical side of the owner's AV operational needs
- IT Representative — to address issues in utilizing or accessing the IT system and how it relates to the user's activities
- Administrative Representatives/Stakeholders — including owner executives, facility managers, department heads and funding organizations

Attendees From the Design Team

- AV Consultant Project Manager or AV Integrator Project Manager
- AV Designer
- Architect — to provide input regarding space allocation and planning
- Program or Construction Manager — who may be on board as a stakeholder in the overall project outcome

PROGRAM MEETING AGENDA

The main objective of AV designers during the program meeting is to gain an understanding of the users' needs in order to develop the Program Document that will include descriptions of the needed systems, along with preliminary budgets. Depending on the scope of the project and number of users to be interviewed, the meeting can take many forms. It may be a single meeting with a few individuals that lasts for a few hours, or it may be a series of meetings with different groups over several days.

Regardless of the size and duration of the meeting, the issues to be addressed are similar. The agenda for the meeting should typically include these items:

Functionality vs. Equipment

All present should understand that the goal of the meeting is to determine functionality, not equipment. While some discussion of equipment is inevitable (especially in short term projects), functionality should be stressed.

For example, projectors often come up in discussion since they are important to the budget and the room. Discussions of projectors in the program meeting should be driven more by what the users want to do, how many images they need to support what they do, what aspect ratio the images should be, what resolutions and sources are to be displayed, and for what purpose. The answer to these questions will then determine the parameters of the projectors that will ultimately be specified.

Technology Trends

Depending on the sophistication of the users, a discussion of trends in AV technology, particularly as they apply to the user's applications, is often helpful as long as it doesn't lead to extravagant "wish lists."

End-User Needs

Usually the facility type is known before the meeting — a new classroom building, distance learning center, corporate training center, headquarters building, boardroom, network operations center or a hearing room renovation. The building or facility type will usually suggest the typical types of AV systems that may be required based on the users' applications.

Examples of an End-User's Needs
- To communicate within a company
- To operate a conference center
- To market or sell a product
- To provide primary and secondary education
- To provide adult education
- To run a local, federal or state government agency
- To monitor and operate an area-wide nationwide or worldwide network
- To provide entertainment in the performing arts, sport venue or themed entertainment venue
- To enhance the worship experience

A word about owner politics

At times relationships within the owner's organization may affect how the program meetings are scheduled, attended and managed. It is important to identify any differences among the end-users, the decision-makers and the funding groups. If they are separate entities, conflicts may emerge between the end-users' needs and/or desires, the decision-maker's perspective, and what the financiers are willing to pay for. If the design team can decipher these relationships before any meetings are held, the meeting leader can be sensitive to potential problems in addressing the issues of each.

Each need may involve a wide variety of applications or activities that would support the user's needs. The important part of the program meeting is to focus on these applications as they relate to the end-user's needs.

Examples of Typical Applications

- General meetings
- Sales force education and coordination
- Product demonstrations
- Classroom teaching and training
- Remote medical diagnosis or training
- Musical or dramatic performance
- Legislative proceedings
- Court proceedings
- Board meetings
- Executive briefings
- Information display

AV Tasks and Parameters

Once the required applications and activities have been captured, the AV tasks that support them can be investigated in more detail. This information determines the overall system configurations and budgets documented in the Program Report.

There are two levels to the information needed at this point:
1. Identification of the tasks that support the applications
2. Identification of the parameters of each task

At the task level, the AV team needs to know what type of AV functions and systems will be needed to support the applications. For example, a common AV task to support many applications is the display of images. If this is needed, then the result of the programming process for this task is establishment of the parameters of the image display, the number of images needed at one time, the resolution of the images to be displayed, the type of video sources to be displayed, and the aspect ratio of the images. These parameters feed into the design process to determine both the facility requirements (display size, room size, room configuration, and lighting, for instance) and the system requirements (e.g., type and brightness of projector, number of inputs and video switching requirements).

Two examples of tasks that may require that parameters be determined are shown in the table below:

AV Task	Parameters of AV Task
Image display	<ul><li>Number of simultaneous images</li><li>Source resolutions</li><li>Sources/signal types to be displayed</li><li>Aspect ratio of sources</li></ul>
Audio playback	<ul><li>Number of audio sources</li><li>Audio signal types</li><li>Area to be covered by loudspeakers</li><li>Distribution of audio to other locations</li></ul>

These and more AV tasks that may need to be addressed are included in a list at the end of the chapter.

Infrastructure Questions

Infrastructure issues must also be addressed to obtain information on the impact that the AV systems will have on the space or building design. Some of the areas that should be covered and questions related to each area are listed below:

Space Allocation

If spaces have already been allocated, part of the AV program exercise is to verify that those allocations will accommodate the application. Questions might include:

- Is there adequate seating and workspace area?
- Is there adequate accommodation of a presenter and lectern or podium, if needed?
- Is there adequate space for equipment including rear projection, control or equipment rooms, if needed?

Chapter 4 — The Program Phase

- What are the sizes and locations of the required images, and can they be accommodated?
- What issues related to acoustics, lighting and HVAC might have an impact on the overall project budget and space allocations?

Engineering and infrastructure questions that should also be addressed include:

HVAC and Security
- What are the hours of operation for each section of the facilities that will house the AV systems?

Electrical/Lighting
- Is a lighting dimming system required?
- What areas require zoned lighting?
- Are control interfaces required for lighting, drapes, shades or other systems that are not included under the AV scope of work?
- Are there non-standard lighting systems required (e.g., for videoconferencing or performing arts functions)?
- Will there be additional or special electrical systems dedicated to the AV systems?

Data/Telecom
- Are special or additional data/telecom outlets and services required for audio, video or control systems?

Step Four: Write the Program Report

At the conclusion of the program meetings, the information is captured in a written report of the findings, including an interpretation of the users' needs with respect to the AV systems. The report should include a conceptual/functional system description along with any needed information about its impact on spaces that have already been programmed, designed or built. Specifically, the program report should typically consist of the following:

- Executive summary
- Systems descriptions
- Infrastructure considerations
- Special issues
- Preliminary AV budgets and terms
- Breakdown of probable costs
- Additional costs (taxes, markups and contingencies)
- Operational staff expertise level required
- Maintenance budget requirements and life cycle expectations

The objectives of the program report are to:

- Communicate to the decision-makers about the overall systems and the budget
- Communicate to the users the system configurations that would serve the needs identified during the program meetings
- Communicate to the design team a general description of the AV systems and what impact they may have on the other trades
- Communicate to the AV designer the scope and functionality of the AV systems to be designed and installed

This document is not an equipment list proposal; it is a functional description.

That said, in some cases the report can be augmented with an equipment list and pricing, but it should always include a justification and description of the systems (in layman's terms). Small and fast track design-build projects are most likely to use a report that is really a combination of a program and a proposal. A proposal that only includes equipment to be installed, however, should never be considered a program report.

While a complete report should include elements that meet the objectives noted here, the following is a recommended report structure:

Executive Summary

The program report is usually read by a wide range of stakeholders in the project. Stakeholders may include a CEO, CIO, architect, AV manager, end-user, electrical engineer, mechanical engineer or facilities department manager.

An executive summary at the beginning of a report provides a brief overview of the entire document. It highlights, in one to three pages, the gist of the report in terms of findings from the interviews, systems to be designed, and a reference to the overall budget required for the systems. This summary serves everyone and is especially beneficial for projects involving large and/or complex AV systems.

Elements to be addressed in the executive summary

- **An overview of the project** — including what the facility is, how it serves the owner, and why it is being built, upgraded or renovated
- **An overview of the programming process** — including who was interviewed at the departmental or organizational level or other individuals who were important during the information-gathering step
- **A short overview of the systems** — identifying what and where the systems are, their overall type and quality, and which users they serve
- **A mention of any special issues** — including any particular concerns of the owner or the project team, e.g., financial issues that could leave the systems budget under-funded, or physical problems such as inadequate equipment or insufficient control room space in an existing plan or facility
- **A reference to the overall budget** — which could be the actual figure or a reference to the budget section of the report

Systems Descriptions

The heart of the report contains a description of each of the systems to be provided. Descriptions may be presented either in a narrative form, with supporting graphics representing the systems, or in a graphic manner with supporting text to clarify and augment the graphics. In either case, the combination of graphics and narrative makes the document more interesting and understandable to the widest range of readers.

Depending on the project size, the number of systems and the different user-types or entities involved, the system descriptions can be organized by space, by user, by system type or a combination of all of these.

There may be systems that are specific to a single space or that appear in several similar spaces that span a number of departments. There may also be facility-wide systems that connect some or all of the AV systems together, or provide overall control, monitoring, interconnection or help desk functions to a building or campus. The nature of the project will determine how the system descriptions should best be organized. Outlines showing three options for organizing descriptions are provided in the sidebar.

Program report structure

Description order based on Area, Entity or Group
Spaces for each group are described under separate headings for spaces that are mostly unique

Department 1
[description]
- Classroom
  [description]
- Videoconferencing Room
  [description]

Department 2
[description]
- Classroom
  [description]
- Auditorium
  [description]

Description order by Space Type, Multiple Groups
If there are several types of spaces that vary from group to group.

Meeting Room
[description]
- Department 1
  [description]
- Department 2
  [description]

Auditorium
- Department 2
  [description]

Description by Space Type with only one Group
If there is just one group.

Basic Classroom
[description]

Advanced Classroom
[description]

Videoconferencing Room
[description]

Auditorium
[description]

If room types and descriptions can be standardized, then for each of the structures above there can be graphical and/or text descriptions for each similar space type that are referenced in a matrix to show which department gets how many of each space type.

Discussion of Infrastructure

General discussion on infrastructure considerations for the project should be included in the report. AV systems often have a significant impact on the architectural, mechanical and electrical systems. Awareness of where the impact may occur is critical to a complete understanding of the project. The areas that should be addressed are:
- Lighting
- Electrical
- Mechanical (both noise and heat)
- Acoustical
- Data/Telecom
- Structural
- Architectural (space plans, adjacencies, allocations, and other architectural issues)
- Interiors (finish requirements)
- Coordination of trades with AV
- Budget impacts of all of the above

Discussion of Special Issues

A discussion of topics requiring special attention may need to be included in the report. Examples are major project obstacles or limitations, project schedule issues, options for specific spaces or overall system configuration options. General discussion of infrastructure budgets can be included under the infrastructure discussions, but there may be more space- or trade-specific issues that require additional discussion or illumination.

Assessing Preliminary AV Budget and Terms

The AV program document should include an opinion, or estimate, of how much the systems will cost the owner. This opinion establishes a more accurate cost for the AV portion of the overall project budget. Terms used when discussing the potential cost of AV systems, or other costs in the project, are as follows:

Opinion of Probable Cost

This term has come into use in recent years to describe an early attempt to determine the cost of a system before there is enough detailed design to produce a line-item estimate.
The opinion of probable cost is an "educated guess" based on experience and some line-item costs for large equipment such as projectors or large matrix switchers. Final costs cannot be applied until the system is designed and the actual equipment selected.

Estimate

An estimate implies that there is a more objective basis for the cost provided. It is an "approximate calculation" that includes a line-item analysis for equipment and labor (perhaps including taxes and other ancillary costs); it would be more accurate that an opinion of probable cost.

Quote

A quote is a detailed and enforceable estimate. It should be provided and identified as such for an AV system only if it is based on a program report and a systems design. If it is not, then the owner, end-users, and the AV provider are at risk of a painful mismatch between the end-users' needs, the system capability and the system cost.

Making the best use of the program report

As added value, the report can become a vehicle for conveying acoustical and lighting criteria and identifying the presence and quantity of architecturally integrated AV components, such as projection screens and blackout shades for each space.

If the AV program report is developed early in the project process, these specifics can be addressed in the design phase. If AV programming is occurring during the architectural design development phase or later, however, it may be crucial to include more specific infrastructure information in the program report.

Budget

Although the term "budget" is often used in the context of the terms described above, by strict definition it applies only to what the owner or project team has allocated for a particular system, trade, or for the whole facility.

The correct relationship of these terms is that the budget should be established based on an opinion of probable cost or an estimate. A quote is then submitted by a provider based on an RFP. The quote is subsequently compared to the budget before acceptance by the owner.

Breakdown of the Costs

Costs are typically organized in the same manner as the system descriptions unless the owner or architect requests a specific budget breakdown. A basic breakdown may be reported simply by room or system (with no further cost detail) and a total.

The costs may be broken down further into subsystems and major equipment, such as projectors, audio systems, video processing systems, large switchers, control systems, and programming and signal distribution systems between rooms. Any breakdown beyond this level of detail can be misleading to those unfamiliar with the design and construction process. It is also likely to be inaccurate — at least at the detail level — later in the project, especially if final equipment choices are made a year or more later.

However, as noted earlier, small or short-term projects in a design-build process may allow for a more detailed estimate or quote to be developed in conjunction with the program report. Such an estimate or quote should include a caveat stating that revisions may be needed if there are changes in the report (and hence the scope or configuration of the AV systems) after being reviewed by the owner. A time limit on the validity of the quote may also be included.

Additional Costs

In all cases, the costs would include equipment plus the labor to install the equipment. Sometimes, costs are broken down to show overhead and profit figures, but this is usually not the case for an opinion of probable cost. There are other costs, however, that should be considered and stated as separate line items, such as taxes, markups, and contingencies.

Taxes

Some owners are exempt from sales tax on all or part of equipment and labor purchases. This should be explicitly noted in the text and shown in the cost breakdown so that it is clear if taxes are included and what percentage is being applied.

Markups

Depending on the anticipated contract arrangement, markups may need to be included. These are usually associated with the subcontracting process. For instance, if the AV systems and labor costs are contracted to a general contractor, he or she may charge the owner an additional five to ten percent or more over the AV systems cost. Clearly, this markup may be significant depending on the cost of the AV system. A two million dollar AV system, for example, might have a markup of $100,000 to $200,000 applied to it.

Contingencies

While not a common practice in the AV industry at this time, contingency budgets can be an important consideration in larger, longer-term projects (those costing over $500,000 and lasting for one year or more). These projects are more prone to potential site conflicts, equipment updates, or new technologies that require additional funding. A contingency budget of five to ten percent of the AV systems' total cost can cover unanticipated overages. Sometimes, the building budget has a contingency that will indirectly cover overages in AV. The most undesirable situation is one in which the AV system is being paid for out of the contingency, which leads to less budget control and accountability.

Step Five: Distribute the Report

The completed report should be distributed to all parties for review and signoff. A set period of time for the recipients to meet, make the necessary decisions and provide comments should be established.

BEST PRACTICE

Don't shop the program document

On some projects, general contractors may be tempted to "shop" the program document with potential AV bidders to obtain review comments and preliminary pricing. This is often done with good intentions, but without an understanding of the AV design and bidding process. The program report is not a bid document and should not be considered as such. It may also include anticipated system costs that would be detrimental to a future bid process. The general contractor should be informed about the purpose of the program document and agree not to distribute it inappropriately.

The report may need to go through revisions and be re-issued to obtain formal approval from the owner or a representative. The following individuals are generally included in this final stage:

The Owner

The owner will review the suggested systems design approaches, the users' requests and pricing. Ultimately, the owner needs to agree and sign off on the systems being provided and on the cost for those systems. Keep in mind that the owner's organization may not include the actual end-users.

End-Users/Technology Managers and/or AV/IT Technology Managers

Having provided information for the program development, the end-users now need to review the document for accuracy. Are their needs being addressed? What are the AV systems concepts, approach and costs? By reviewing and signing off on the program document, this group establishes its agreement with the scope and purpose of the AV systems.

Architect

The architect will be interested in the scope of the AV systems and what impact they may have on the facility design and the project budget. The architect does not usually sign off on the program — that approval is reserved for the owner and end-users — but, if the AV team is subcontracted to the architect, he or she often serves as the clearinghouse for distributing the report and collecting feedback. The architect would also distribute the report to the consulting engineer (MEP) in high-end projects.

Construction Manager

The construction or program manager, if there is one, will have a tremendous interest in the program document, since it will be the CM's guide to manage the AV systems integration within the overall scope of the facility project. In some cases, the CM or PM may even be a part of the sign-off process along with, or representing, the owner.

General Contractor

Sometimes, a general contractor may be on board during the design phase to assist with project costing or as a part of a fast-track project. In such cases, the architect may be providing construction documents to the GC in stages, without completing the entire design package first. If so, the general contractor might be privy to the program document in order to be aware of activities during the program phase and upcoming issues in the architectural design.

AV Integrator

If the AV integrator is preparing the AV program report as part of a design-build project, then the internal team should be included in the report distribution for their comments and education. The integration team should also be included in the distribution if the project is under a consultant-led design-build process.

If the project is to be design-bid-build, however, then it is unnecessary (even inappropriate) to provide the program report to an integrator at this stage. The program report (not including cost information) can be provided to potential bidders later in the process as a component of the RFP (see Chapter 5).

Cost Estimator

Professional cost estimators assist in managing the budgeting process but often do not have the expertise to accurately develop costs for pro-AV systems. Once they have been developed, the estimator can incorporate the AV costs into the overall project budget.

Other Recipients

In some projects, other professionals with special interests should receive the report. For example, a funding organization that is donating money for a facility, but is not directly involved in the project design and construction, may have a need for a copy. The owner may specifically request distribution to other individuals. These additional distributions are acceptable, as long as the recipients are aware of the purpose of the program report, their responsibility for a response if required, and the appropriate limitations on redistribution of the report.

Step Six: Approve the Program Report as a Basis for Systems Design

Once the review process is completed (perhaps with additional follow-up meetings), the program report should be formally approved by the owner and end-users. Upon approval, the document can be used as a basis for the design of the AV systems and their infrastructure. The Basis of Design document protects the owner and consultant by ensuring that everyone is defining the project in the same way before more money is spent. At this point, one of three paths can be taken depending on the project process being used:

The Underpinning of a Design-Bid-Build Process

If the project is proceeding as a design-bid-build process (typically led by an independent AV consultant), the program report becomes the basis for the design of the AV systems. During the design process, the report is used to guide the AV designer in preparing the bid package to be distributed to potential integrators later in the project.

If the end-user's needs were not sufficiently defined before the program phase, the contract for design services may not have been established or proposed. In this case, the program should also be the basis for the AV consultant's proposal for the continuation of design and construction administration services.

Continuing a Design-Build Process

If an integrator had been contracted for a design-build project before completion of the AV program report, the contract may have been established based on an assumed scope and pre-determined AV budget. In this case, the AV program document defines more fully what the system design and installation will be. Modifications to the integrator's original contract may be required if the program results differ from what the integrator first proposed.
If, as in the design-bid-build scenario, the project scope was not sufficiently well-defined before programming, the integrator should develop a proposal for continuing design and installation of the systems based on the program report.

Incorporating the Report into a Design-Build RFP

In some circumstances, the program report may be used as the basis for a request for proposal (RFP) to move into a design-build process. Two common scenarios may drive this decision:

1. The owner has completed an internal program development process and wishes to pursue a design-build contract with an integrator.
2. A fast-track project has been programmed by an independent AV consultant and there is insufficient time for the sequential design-bid-build process. The owner and/or consultant would then manage the contracting process and manage or monitor the design-build process once an integrator is engaged (establishing a variation of the consultant-led design-build option).

Chapter 4 Checklists

AV TASK PARAMETERS FOR DEVELOPMENT OF THE AV PROGRAM

The following table is a list of possible AV tasks that may be required to support the end-users' applications and needs. It is this level of information that is ultimately needed to create the AV Program Report and its opinion of probable cost for the AV systems.

AV Task	Parameters of AV Task
Image display	• Number of simultaneous images • Source resolutions • Sources/signal types to be displayed • Aspect ratio of sources
Audio playback	• Number of audio sources • Audio signal types • Area to be covered by loudspeakers • Distribution of audio to other locations
Speech reinforcement	• Number of talkers to be reinforced • Location of talkers • Area to be covered by loudspeakers • Interface to other systems
Audioconferencing	• Stand-alone or part of videoconferencing • Concurrent with speech reinforcement? • Local bridging for multi-party audio
Videoconferencing	• Dedicated function or incorporated with presentation system? • Number of participants • Single axis or dual axis (participants in audience only or is there also a presenter?) • Number of images required • Resolution of conferencing images • Type of connections to be supported (ISDN, IP, satellite, fiber, broadcast, etc.)
Overflow / Interconnection of spaces	• Identification of potential source spaces • Identification of potential destination spaces • Number, type, resolution and format of audio and video signals to be connected • One or two way connections required? • Are sites within or outside of the project facility? • Type of connections to be supported (ISDN, IP, satellite, fiber, broadcast, etc.)
Recorded media playback	• Types of recorded media • Audio and video parameters of media content • Accessed by technicians and end-users?
Recording of events	• Audio and/or video • Audio and video parameters of recorded signals • Set-up and controlled by technicians and end-users?
System Control	• Locations of control within each space • What do the control systems need to do? • Who will be controlling the system (end-users, assistants, technicians)? • Local and/or remote control required? • System-wide monitoring • Help desk functions • Interfacing to other devices or systems required (lighting, drapes, building automation system, etc.)?
Ancillary Systems	• Are there any ancillary systems required such as audience response systems, background music, sound masking, nurse call, security or other systems that are needed to support the users' activities or spaces?

Overview

A. Project Overview
   1. Identify project type
   2. Project schedule
   3. Identify major project applications required for end users' operations
   4. Identify user groups and functions

B. Identification of owner and end-user vision and style

C. Differentiation of AV system functionality vs. AV equipment
   1. Define expectations of the process
      - What is needed/expected from the design team?
      - What is needed/expected from the owner/user?
      - How will the design team and owner/user communicate?
      - System quality

D. Technology Trends

Review Existing Documents, Facilities and Infrastructure

A. Review pertinent parts of architectural program
B. Review any existing AV program information
C. Tour existing facilities

Identify User Functions

A. Existing functions
B. Anticipated functions

Identify Overall User Standards and Requirements

A. Standards
B. Benchmarks
C. Known connectivity requirements
D. Known basic audio-video requirements
E. Internal tech support availability
F. ADA and section 508 issues

Discuss Each Space or Area

A. Identify each area requiring systems

B. Space-by-space functional review
   1. Functions required for each space
   2. Operational requirements (day, night, remote monitoring)

C. Identification of AV tasks and parameters for each area
   1. Identify major equipment requirements (number of images required, room size and seating, conferencing required, audio and video sources)
   2. Identify potential impact on infrastructure
      - HVAC
      - Security
      - Electrical
      - Lighting
      - Data/telecom

D. Owner furnished equipment

E. Budget issues and priorities

Conclusion

A. Identify key individuals and contact information for follow-up
B. Identify follow-up meetings
C. Discuss schedule for completion and distribution of report

---

AV NEEDS ANALYSIS/ PROGRAM MEETING AGENDA SAMPLE

The AV Program Meeting may include some or all of the items for discussion as noted in this sample agenda. The process may require several meetings to interact with different stakeholder and end-user groups. This agenda should be used in conjuction with the sample AV Program Questions included in the Appendix.

BENCHMARKING CHECKLIST

The two typical benchmarking opportunities during a building and system delivery process occur during the program phase and the design phase. Once the benchmarking tour is set up, this list can be used to help make the most of the visits and glean the appropriate information that will be useful to the design team, the owner and the end-users.

Facility Information
- ❏ Organization, facility and location
- ❏ Contacts at the benchmark organization
- ❏ AV project installation date
- ❏ AV budget at time of installation
- ❏ Delivery method used for the project
- ❏ Style of the project – low-, mid- or high-end
- ❏ Is there any facility or system documentation that can be shared with the design team?
- ❏ Is photography allowed?

Benchmark Information
- ❏ What is the benchmark facility's focus?
- ❏ How is it like the facility being designed?
- ❏ How is it different?
- ❏ What technologies are used and how do they support the end-users' activities?

With respect to the project being designed, what features are of particular interest at the benchmark site with regard to:
- ❏ Audio systems
- ❏ Video systems
- ❏ Control systems
- ❏ Integration with IT
- ❏ Lighting
- ❏ Acoustics
- ❏ What technologies, design approaches or criteria were used at the benchmark facility?
- ❏ What are the benefits and drawbacks of their design approach?
- ❏ What accommodations were made for upgrades and additions to the systems?

Facility Management
- ❏ How does the benchmark facility organization manage its AV technology?
- ❏ How is the benchmark project serviced in terms of operations, maintenance, help desk?
- ❏ How much staff is required to operate and manage the AV systems and what are their qualifications?
- ❏ What are the benefits or drawbacks of their technology management approach?
- ❏ How do these approaches compare to the existing and/or planned facilities?

Owner and User Feedback
- ❏ What do the benchmark facility end-users have to say about the technology and the facility?
- ❏ What do the benchmark facility technology managers and technicians have to say about it?
- ❏ What do the benchmark administrators have to say about it?

Conclusions
- ❏ What features of the benchmark facility should be included in the project design?
- ❏ What features of the benchmark facility should be avoided in the project design?
- ❏ What aspects of the benchmark facility's technology management approach should be developed or avoided in the current project owner's organization, and what impact does that have on the facility and systems?

CHAPTER 5
The Design Phase

Step One: Start the Design Phase with a Kick-Off Meeting
Step Two: Establish the Infrastructure
Step Three: Developing the AV System Design Package
Step Six: Making the Most of Review

CHAPTER FIVE

Design — the art of translating what the end-user needs into documents that show how it is to be built. The process of audiovisual design may be conducted during the overall design phase of a new building, or independently as an AV-driven process for the renovation of an existing facility. In either case, the necessary design documents are produced to communicate to the AV installation team what the owner needs (and wants) in terms of systems and infrastructure.

This chapter covers the processes for creating both the infrastructure design and the AV systems electronics design.

Step One: Start the AV Design Phase with a Kick-Off Meeting

The design process described here includes all tasks that are typically included in a building design — even on smaller projects where some steps may be consolidated.

The infrastructure design phase typically begins with a contract that also includes the system design. By way of formal kick-off, a meeting is called to establish the logistical parameters and set expectations for the design and integration process. Certain members of the project team should participate as follows:

Recommended Kick-Off Meeting Participants

- The architect
- Construction/program manager
- Owner's administrative/construction representative
- AV provider project manager
- Electrical consultant
- Acoustical consultant
- Data/telecom consultant
- Mechanical consultant
- End-user representative
- Owner's AV technical representative
- AV system designer (if different from the project manager)

Some projects may require only a few of these participants in the design kick-off meeting, while others (particularly large AV-centered projects) may require all of these participants and others.

SETTING THE KICK-OFF MEETING AGENDA

This meeting is critical as an opportunity to establish good communications and to ensure that all parties are aware of the AV design process requirements. It should be kept streamlined and focused. Typically, these meetings are most effective when the agenda contains the following discussions:

AV Infrastructure Design Coordination

There will be several milestones that should be met during the architectural/engineering design phase. This will entail an interactive discussion about primary requirements, a timeline, coordination, and who will provide which elements. The primary discussion items are floor plan and architectural coordination, electrical power and conduit requirements, data/telecom requirements, lighting design requirements, acoustical issues and mechanical systems coordination.

AV System Design Coordination

Depending on the project schedule and size, the system electronics design may take place either during the course of the infrastructure design or upon its completion. Discussion should focus on design review periods and submittals, the bid date (if any), the system integration start date and facility occupancy date.

AV Installation Scheduling

System installation will not take place during the design phase. Consequently, while it is useful during the kick-off meeting to touch upon the schedule for installation, the details of the installation schedule may not have to be set at this point.

Occupancy Overlap

It is useful to discuss the potential of AV installation overlapping with end-user occupancy. Such an overlap can hamper the AV systems installation because dust-free conditions are critical to successful installation of much of the equipment. Equipment failures due to dust can have a negative impact on the end-users' first experience of the system, or worse, may cause long-term damage to equipment which will not be covered by manufacturers' warranties. AV installation occurring after move-in also may have a significant impact on the users' scheduling of their operations during that period.

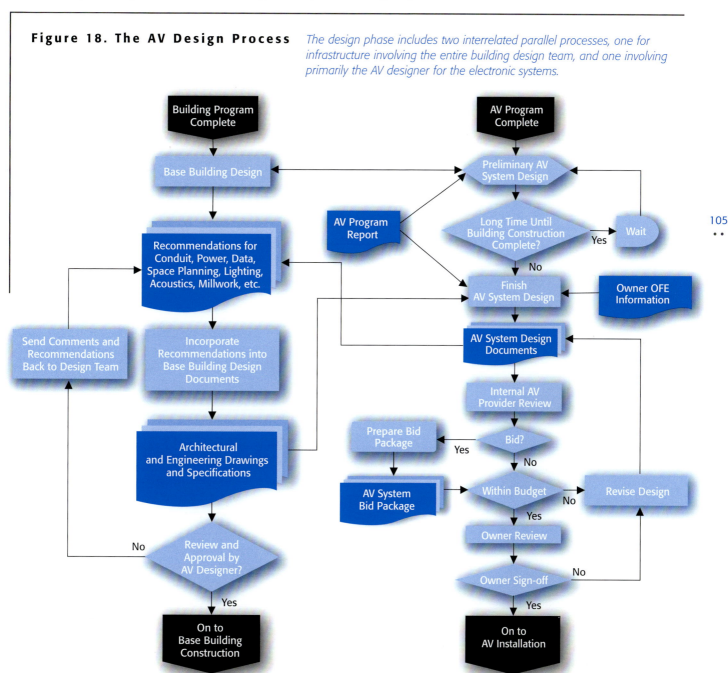

Figure 18. The AV Design Process *The design phase includes two interrelated parallel processes, one for infrastructure involving the entire building design team, and one involving primarily the AV designer for the electronic systems.*

Chapter 5 — The Design Phase

Carefully evaluate OFE equipment

Supplied equipment can work if it is evaluated for acceptable performance. The age, type and brands of the technology or equipment need to be taken into account. It is critical to assess whether or not the equipment will still be serviceable when the project comes to fruition as well as far beyond.

AV Procurement Methods

Procurement methods should be discussed because these will have an impact on how the design and construction documents should be prepared. If a bid process is used, then bid contract documents will be required. If a design-build contract is in place, then the system design documents are needed only for review by the owner and end-users. However, the infrastructure design process should be the same under any procurement method.

Owner Coordination

Any owner or end-user requirements related to AV must be defined up-front so that appropriate documentation can be provided to the AV designer for incorporation into the design. Requirements may include standards for equipment manufacturers, touch screen interfaces, AV furniture, lighting, acoustics or system configurations.

OFE (Owner Furnished Equipment)

If the owner and/or end-users want to use existing or separately purchased equipment (also known as CFE or customer furnished equipment), much coordination and discussion should occur. Re-use or purchase of this equipment (including computers, AV furniture and audiovisual equipment) often raises compatibility and warranty issues that the AV designer needs to address in system design and bid document preparation. This is also an opportunity for clarifying for the design team and the owners any expectations about who provides what equipment and under what contract.

AV Budgets and Fees — Opinion of Probable Cost

The meeting also provides an opportunity to confirm the project budget for the AV systems and any alternates that may be needed in the system package. Typically, an opinion of probable cost, based on the project team's budget for the AV systems, should have been developed during the program phase. If it was not developed, this is the time to discuss the AV budget and how it may affect the final system configurations.

Project Confidentiality

Some projects may be confidential in nature — either because the facility is secure on an ongoing basis, or because the owner wants the details (or existence) of the project to remain confidential until the facility is open. Often cited in the design or integration contract, there should be a discussion of the reasoning and logistics of maintaining confidentiality.

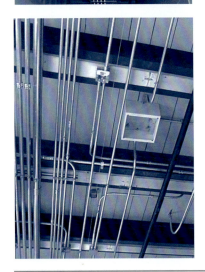

Step Two: Establish the Infrastructure

It is important to understand that the infrastructure is based on the design of the electronic systems. Yet, in many longer-term projects, the final systems designs are not completed until after the infrastructure design is done. In shorter-term and fast-track projects, particularly in design-build delivery methods, the systems and infrastructure designs may be conducted simultaneously.

The infrastructure design is partly focused on architectural elements (such as space plan, screen locations and seating arrangements) that are being developed by the architect. For longer term projects, the electrical and mechanical requirements that are more closely tied to the AV electronics and wiring (such as power, conduit, back boxes and data/telecom), are based on conceptual designs for the AV systems and past experience of the AV designer.

Provide infrastructure design for all potential AV systems in the program

In some cases, the programming process may result in systems that are beyond the available funds to install. If this is the case, it is best to provide the infrastructure to support the complete system — even if the initial systems installation does not include the full program. This increases flexible options for future planning.

Infrastructure is designed throughout the architectural design phases as outlined below:

Conceptual Design

To help the architect arrange the facility's spaces with AV in mind, the AV designer needs to provide input to the architectural team on a number of issues, including:

- Mechanical and electrical room placement based on acoustical noise control issues
- Overall room sizes based on room function
- Size and relationship of projection and control rooms (as needed)
- Sightline issues that may affect ceiling heights and ultimately structure-to-structure heights
- Acoustic and lighting criteria if they have not already been established during programming
- Location of cameras and view angles

Schematic Design

Additional aspects of the facility need to be addressed as the space plans become more established. If the architect moves into schematic design without a separate conceptual design, the items above also need to be addressed. During schematic design, attention to the following items is needed:

- More attention to and refinement of the space relationships and sizes
- Clarification of acoustical and lighting requirements
- Evaluation of potential cable pathways
- Ceiling heights
- Equipment room requirements
- Room orientations
- Structural issues related to AV equipment mounting or sightline requirements

Design Development

At this point in the process, more of the engineering coordination can begin. For the AV part of the project, the following tasks should be accomplished:

- Conduit and back box location plans are developed
- Data outlet requirements provided to data/telecom consultant
- Power requirements provided to electrical consultant
- Heat loads provided to HVAC consultant

Figure 19: The Sightline Study

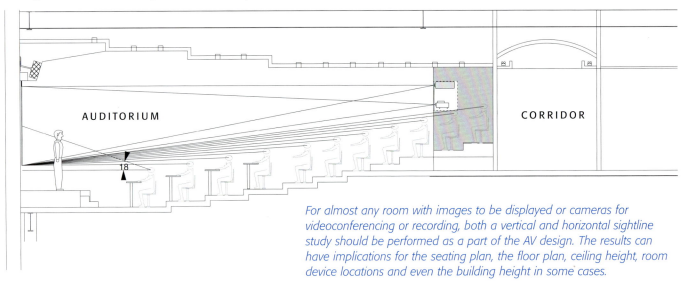

For almost any room with images to be displayed or cameras for videoconferencing or recording, both a vertical and horizontal sightline study should be performed as a part of the AV design. The results can have implications for the seating plan, the floor plan, ceiling height, room device locations and even the building height in some cases.

- Detailed sightline studies conducted to verify room designs
- Lighting designs initiated
- Develop acoustical architectural designs
- HVAC and electrical system noise control issues reviewed as HVAC and electrical systems are developed
- Examination of surrounding environment to check airborne and structure-borne noise and vibration from people and equipment in adjacent spaces
- Detailed room finish requirements where needed (particularly for rooms in which videoconferencing will occur)
- Check wall elevations so that projection screen locations, rear screen projection screen openings and AV device locations can be established and coordinated.
- Mounting requirements and room layouts prepared. Coordinate structural loading information and mounting details.

Construction Documents

This is the phase where some of the most intense design coordination occurs. AV design issues that need to be addressed during this phase include:

- Finalization of conduit, cable tray and back box sizes, pathways and locations
- Review of room lighting and lighting control designs for AV spaces
- Detailed review of acoustical design issues, including mechanical, electrical and architectural acoustical design elements
- Review and finalization of AV design details such as structural mounting, wall elevations, screen type, size and locations, AV-related room configurations and other details that are critical to owner and end-user satisfaction.

VALUE ENGINEERING AND COST CUTTING

If the project cost estimate exceeds the project budget, or if the project's funds have been reduced, a process called "value engineering" is sometimes undertaken to lessen costs while maintaining the original scope and quality of the project.

If the quality and scope cannot be maintained within the current available funds, then scope and/or quality reductions are considered. At this point, it is no longer value engineering, but simply cutting costs. It is important to distinguish between the two during any effort to reduce cost. Both can reduce costs, but one does it while maintaining scope, function and quality, the other does not.

For the base building, quality reductions may include lower cost finishes (such as changing a wall from marble or wood to vinyl wall covering or paint), using less expensive light fixtures or changing coffered ceilings to lay in acoustical tile. Scope reductions would go further and look at ways to reduce floor area or simply build an unfinished "shell" in certain areas that would enclose, but not contain interior walls or finished ceilings.

These types of cost reduction have the potential to affect the AV systems in many ways. For instance, reduction in the numbers or quality of light fixtures or deletion of dimming systems may cause problems in AV presentation spaces. Deleting acoustical wall paneling or door seals, or using simpler and lighter wall constructions can have a negative effect on the acoustics of AV spaces. All of these issues should be carefully monitored by the AV design team during any value engineering process.

Value engineering will inevitably involve the AV systems themselves. In an attempt to maintain scope and quality, the AV designer should ask questions such as:

What are the alternative approaches and substitutions?
Can the system be designed with different technologies, equipment or configurations that reduce cost, while maintaining the AV program scope and quality? For instance, can designing a different AV switcher configuration save money?

Are there subsystem deletions?
Are there subsystems within a room's system design that can be deferred without impacting other system functions? An example would be the removal of desktop microphones, their associated mixers and echo cancellers for built-in videoconferencing or for distance education while ensuring that the presenters' speech reinforcement and all intended AV presentation capabilities are not compromised.

Are there component reductions?
Can the number of items be reduced? For example, by using one projector instead of two in a dual-presentation room, or by sharing a portable document camera. While future wiring and other components should be considered for future ease of installation, there is potential here for major costs to be eliminated.

Are there system deletions?
Are there shell spaces that will no longer require AV systems? Are there some rooms that will be finished out and used that do not currently require the AV systems, in which case some equipment and labor can be deleted from the current scope and addressed later as funds become available?

Value engineering and how it affects contracts

The value engineering process can substantially change the basis of the systems and building design.

This may require revising existing documents, which, in turn, may have an impact on the negotiated design contracts.

If designers completed their work within the original budget, there could be additional design charges incurred if they are asked to redesign the project based on budget cuts. On the other hand, if the re-design occurs as a result of the designer's original plan coming in over budget, there should not be additional charges.

If the design of the infrastructure and/or system is in progress when the reduction in scope occurs, then the designer may be asked to reduce his or her design fees accordingly.

If the AV systems to be installed have been cut under a design-build contract that is already established, then a contract modification would also be required in this case.

Step Three: Developing the AV System Design Package

With the infrastructure design and installation underway, the design of the actual AV systems takes place. The purpose of the design phase for the AV systems is the same as for the infrastructure and building design: to communicate the design intent from the design team to the installation team so that the system can be constructed and installed. This is true for any of the design methods chosen. The mechanics of the process may vary, but the tasks remain the same across all the options.

COMMUNICATING DESIGN INTENT

The design intent, or how the project will ultimately "look," can be viewed from three separate, but entirely integrated, perspectives — 1) fabrication and installation; 2) functional; and 3) quality.

Fabrication and Installation Intent

Fabrication and installation intent relates to conveying the intent of the designer with respect to the selection and procurement of equipment, cable and other materials, in addition to the fabrication, installation and software programming tasks. This intent is the basis for the majority of the content in the design drawings and specifications.

Functional Intent

While the details of fabrication and installation are essential to the design deliverables, this information must be complemented by the functional information about the systems. This is particularly important within consultant-led design-bid-build scenarios, but also applies to the other methods. The installation team cannot be expected to discern the entire functional intent from installation documents alone. The functional intent must be written in the form of an understandable narrative or a list with accompanying graphics. This is especially true for the control system or digital signal processing (DSP) equipment programming tasks, which require full explanation of the functions that will be written in the code.

Quality intent

Quality is intended for every aspect of the system — from the equipment to be procured, to the actual installation (such as wiring, terminations, rack assembly and signal quality) and to the optimal functionality for end-users. In fact, the higher the quality of the design process[18], the higher the quality of the design phase deliverables, which will ultimately improve the potential quality of the bids (if any) and the installation itself.

RESEARCH AND BENCHMARKING AS A PART OF DESIGN

For many day-to-day AV projects, most design professionals and owner representatives are familiar with the design intent, what the systems will generally "look like" and how they will operate. Standard boardroom, training room, classroom, conference room and hotel ballroom AV systems are relatively common and similar in nature.

However, some projects "push the envelope." Such a project might be a new and unique application, or an unusual project with which many professionals are unfamiliar, such as high-tech aquariums or specialized museum exhibits.

---

[18] Besides communicating design and quality intent, the design deliverables must be enforceable under any method. In bid scenarios, the design package becomes a contractual agreement. Therefore, the wording and graphic elements must enforce the terms of the design contract so that the design intent can be maintained. Under a design-build process, the design documents should be enforceable within the design-build organization as a way of maintaining the design intent throughout the design and installation process.

These special projects may require site visits to similar facilities and/or research during the program phase and the design phase.

Over and above site visits, research may involve a number of avenues:
- Visits to a specialized tradeshow related to the user's industry or application
- Discussions with other AV colleagues with experience on a related project
- Discussions with AV equipment manufacturers that may have been involved on a similar project
- Discussions with manufacturers of equipment that will be part of the special project. This is particularly true if the equipment needed in the AV systems connects to specific devices, e.g. medical devices with audio and/or video output.

CREATING THE DESIGN PACKAGE: DRAWINGS AND SPECIFICATIONS

The design package forms most, if not all, of the contract that the AV integrator signs at the project award. The contract is agreed to:
1. Between the AV integrator and the owner directly; or
2. Between the AV integrator and a general contractor (or construction manager) working for the owner.

Because of contract obligations, a sizeable amount of administrative language is required. Many contract disputes arise from a misinterpretation of either the graphical or textual language that comprises the design documents. The design package consists of two basic components that must be created: Drawings and specifications.

Design drawings are the one common and essential element of the design deliverables among almost all the building trades, including AV. The drawings depict (generally in graphic form) how the systems are intended to be configured and constructed.

The other primary element of all AV bid packages is the set of specifications (generally consisting of text) to complement the drawings. These specifications completely describe the equipment to be used and the required installation techniques. While specifications are not always produced for a design-build project, some text documentation should be developed.

Creating Design-Bid-Build Deliverables

The documents for the design-bid-build process apply when a designer is contracted separately from the AV installer. This would also apply to an owner-designed process where the owner creates a full design package in-house.

In the bid process, the design documents must include enough information for the design intent to be conveyed and enough detail so the bidders' responses can be accurately and fairly compared. In addition to the design information on the system, certain administrative contract language will form the basis of the actual installation contract.

For larger projects where the construction phase of the base building is one year or more, the AV package should be issued separately after the base building construction documents are issued. This should be timed to allow for the bid and contract negotiation process as well as the installation and commissioning.

BEST PRACTICE

Consider the impact of time and money on quality

There is a powerful connection between quality AV design and installation and the time plus money that is necessary to attain them. Longer projects are almost always able to better accommodate a tight budget than fast-track projects. In a fast-track project, one can almost count on spending more money to get the job done while maintaining a quality installation, compared to a project schedule with a more comfortable schedule.

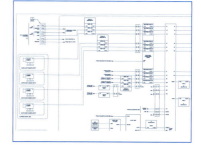

Design-build vs. design-bid-build deliverables

Up to this point in the AV design process, the differences in the deliverables between the alternate delivery methods have not been significant. During this phase, however, more differences surface between the design-bid-build process and the design-build process.

The goal for the documents in each design process is the same (conveying the design intent), but the actual documentation required differs depending on whether or not there is a bid process for the installation.

Major Components of the Design Package[19]
1. Administrative "front end" specifications
2. Architectural and infrastructure drawings
3. AV system drawings
4. AV system specifications

1. The "Front End"

Typically, a construction contract has an administrative front end or general conditions section that delineates how the contractor is to interact with the owner, what documentation is required, meeting attendance requirements, required reports, submittal processes and basic record drawing requirements, and other important details. Components of the front end usually include:

a. **Invitation to Bid.** This is essentially a "cover letter" that formally invites potential bidders to respond to the bid package. It includes the contact information, dates, times and schedules associated with the bid process under those methods that include bidding

b. **Request for Qualifications (RFQ).** If not issued separately, there may be an RFQ included in the front end section

c. **Bid Response Form.** The bid response form organizes responses from potential bidders and indicates what information is required. While there may be drawings, financial statements and other qualifications documents requested as a part of the bid response, the form also contains blank spaces for the bidder to answer specific questions and enter the price for the proposed system. A pricing breakdown may be required as well as a separate list of parts of the system installation that may not be accepted (called alternates).

d. **General Conditions.** The architect for the base building contract generates general conditions. If the AV contract is subcontracted to the general contractor, then the general conditions will apply by default. If the AV contract, however, is let separately from the general contractor, then the general conditions section will often be re-used with little or no modification as the front end for the AV contract documents. General conditions would be used in both a design-bid-build as well as design-build contract. Typical sections of the general conditions include:

- Project work conditions and terms
- Definitions of terminology
- Materials and workmanship
- Safety and accident prevention
- Permits, regulations and taxes
- Insurance
- Bond
- Overall quality control
- Submittals
- Substitutions schedule
- Changes in the scope of work
- Initiating and processing change orders
- Claims and corrections pertaining to the work
- Warranty
- Non-discrimination and affirmative action
- Contract termination options
- Arbitration
- Invoicing and payments

---

[19] Typical package components that are unusual or specific to the AV package. Additional information can be found via resources offered by organizations such as the AIA and CSI. See the end of this chapter for a checklist version of these components.

e. **Money Matters in the Contract.** Within the general conditions and sometimes within a consultant's specifications sections, terms having to do with invoicing and payment are included and are a critical part of the contract. There are a few aspects of these terms that deserve special attention, as noted below.

Communications procedures. The contract documents establish the process by which communications, submittals and approvals are handled. Typically, all submittals and other communications are routed to the owner's project manager and distributed accordingly. If an AV consultant participates in the project, than the integrator should investigate the opportunity to forward an advance or informational copy of all submittals to the consultant. Special circumstances may apply, but they should be known, agreed to, and negotiated prior to scheduling commissioning.

Penalty clauses. When it is critical that a project be completed on time, or if schedule concerns exist that may have a financial or operational impact on the owner, a contract may include a penalty clause (often referred to as liquidated damages). This clause provides for charge-back to the contractor, integrator or others if equipment, services or the completed project are delivered late or in default as defined by the contract.

A penalty clause typically defines the charge-back in terms of cost per time period, e.g., dollars per week beyond a completion date noted in the contract. This can be a fixed date, or a fixed timeframe, commencing with the contract execution. It could also be a milestone relative to project progress, completion or work by others. In all cases, the integrator must maintain documentation citing reasons for delays so that any penalties incurred may be proven to be justified.

Terms and payments. All payment terms must be negotiated and all parties must be in agreement prior to the start of any and all work. For short projects, there may or may not be a deposit required, and the remainder of the contract price is usually remitted at the completion of work. Longer projects are typically invoiced in progress payments, valued at work completed to date. Progress payments normally include, but are not limited to:
- Labor incurred to date, represented as a percentage of the entire project
- Value of goods bought and stored
- Sub-contract work
- Fees, licenses, and others within the terms of the contract

Deposits and mobilization fees. It is often appropriate to pay a mobilization fee or deposit to the contractor to get started on the project. This is particularly important for out-of-town projects. The fee may typically range from 10 to 50 percent of the total contract.

Unlike many construction contracts where materials are delivered to the site, AV integrators take possession of the equipment and assemble it over a period of months at their shop before delivery to the site. Therefore, it is appropriate to reimburse the AV integrator for the purchased equipment (which represents the majority of their costs on most projects). If this is not done, the AV integrator faces a significant financial burden.

The solution is to include a clause in the AV integrator's contract that allows for a bonded warehouse that covers equipment stored at the integrator's shop or includes other special procedures that allow the integrator to bill for items purchased before they are delivered to the site. This may include a requirement for the design consultant or other owner representative to verify that equipment being invoiced is actually at the integrator's shop.

Chapter 5 — The Design Phase

Progress payments. Progress payments include the value of any goods purchased, but not yet delivered, to the owner under the contract. These goods, known as "bought and stored," may be held by the integrator for many reasons, including:
- Inclusion in shop-fabricated systems
- Assurance of quoted pricing
- Security if site is unsecured

Progress payments also include retainage, or a portion of the payments to date which is held back as assurance of project completion. This is typically calculated as 10% of work completed to date, often reduced to 5% after substantial completion.

Invoices that include the value of goods bought and stored should provide proof that said goods have been received by the integrator and are properly stored and insured. For this proof the owner may require some or all of the following:
- Packing slips
- List of serial numbers
- Photographs of goods
- Proof of insurance and/or bonding (may require rider specific to project/owner)
- Physical inspection of goods by owner's representative or consultant

Invoice or application for payment. In construction terms, an invoice or request for a progress payment is known as an Application for Payment. Specific forms to be used for invoicing and the correct process for authorization and payment must be defined in the contract document. Note that the payment process may be different from the ones for submittals and other project communications.

In addition to references to the project, owner, contractor, integrator, and others, Applications for Payment should include:
- Contract value to date (original contract value + net value of approved change orders to date)
- Value of work and goods purchased and delivered or stored to date and retainage
- Previous payments, current payment, and balance to finish

Final payment. Final payment is requested after all contractual obligations have been completed. Several affidavits are normally required, certifying that the work is complete and that all debts and claims against the goods and workmanship delivered within the project scope have been paid or otherwise satisfied. These affidavits verify to the owner that liens may not be made against the installed equipment.

AIA contract forms and templates. The American Institute of Architects (AIA) has established an industry standard for a variety of contract documents. These forms are available for a fee from the AIA (www.aia.org) Some of the more commonly used AIA forms are listed in the appendix.[20]

---

[20] Please note that this list is for informational use only and does not represent the authors' endorsement for use.

2. Architectural and Infrastructure Drawings

Architectural drawings indicating where the systems will be installed, what infrastructure exists and what may be required for AV systems should be included in the bid package. These would indicate available power, signal conduit, data outlets and structural accommodations for AV equipment.

3. AV System Drawings

Drawings that depict the AV system itself should reflect the equipment configurations, interconnections, details, plate layouts and other graphic depictions of the system installation.

Typical components of the design drawings package
- Title and index
- Typical power, grounding and signal wiring details
- Floor and reflected ceiling plans showing device locations
- System functional diagrams
- Rack elevations
- Custom plate and panel details
- Miscellaneous details and elevations including:

    - Speaker aiming information
    - Large scale plans, such as equipment or control room plans
    - Architectural elevations showing AV devices, their location and their relationship to other items on the walls
    - Custom-enclosure or mounting details for projectors, microphones, loudspeakers, media players and others
    - Furniture integration details
    - Any special circumstances or details that may be required for the installers to properly understand the design intent

4. AV System Specifications and CSI Format

Specifications, sometimes called the "project manual," are needed to further describe the system, the components, codes, references and other requirements, in addition to how the system should be installed and tested. Also included is information concerning submittal requirements, shop drawing requirements, component and system testing and any commissioning requirements. Under a design-build scenario, full specifications may not be generated.

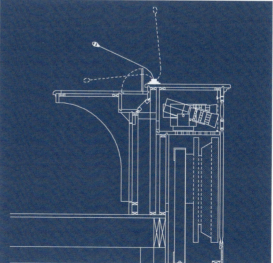

Chapter 5 — The Design Phase

Look for qualified teams

When looking at the qualifications of the potential AV team, individual certification and company commitment make a difference. Those organizations with CAVSP[21] designations and/or InfoComm-certified staff have committed to ongoing education to stay abreast of new developments in the field and to the best practices of the industry.

For almost all construction projects, specifications should be produced in the Construction Specifications Institute (CSI) three-part format. This format applies to each specification section that is created for the entire project. This includes products and systems ranging from HVAC fans and gypsum board to electrical outlets and lighting fixtures, and of course, the AV systems. Specification sections are organized in Divisions that contain descriptions of related work. CSI produces a standardized outline called MasterFormat.[22]

For each section, the CSI specification format organizes the text describing the systems, equipment and their installation in the three parts — 1) General; 2) Equipment; and 3) Execution.

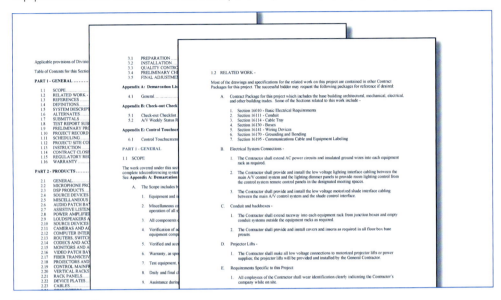

Part 1 - General

The General part of the specifications section includes administrative information that is more specific to the AV contract. This information should conform to the Front End section. Typical subheadings under the General part would include:

Project Description
The design deliverables, whether developed by an independent consultant or an AV designer with a design-build integrator, must be based on the program and must convey the functional intent of the systems. A textual description of the project should be included in the specifications and/or provided as an additional document in order to give the integrator more of the "big picture" concerning the project.

Work Included
These paragraphs provide brief references to what is included in the scope of work for this project and what is not. For example, installation of all AV equipment and cabling may be included in the contract, but conduit, back boxes and cable trays may not be. These general items as well as any peculiar circumstances that may require clarification would be included here.

[21] Certified AudioVisual Solutions Provider, the certified company program offered by the International Communications Industries Association, Inc.® (ICIA®), the international trade association representing the professional AV community worldwide. InfoComm's certification is the oldest, most widely utilized, and most specific program for AV and information technology systems design and installation, and its Certified Technology Specialist (CTS) designation has been officially recognized by the National Certification Commission (NCC).

[22] In 2004, CSI published a major update to the specifications structure called MasterFormat 2004 that placed audiovisual (AV) work under Division 27 along with other communications technologies. Information on the MasterFormat can be obtained from CSI at its website, www.csinet.org.

Alternates
Descriptions of any alternates that may be required for separate pricing. These may be individual rooms or systems, or they may be specific component substitutions or configurations that are to be treated separately so the owner can choose specific alternatives to the base system. There may be restrictions on how the alternates are selected and these should be noted in the specification text.

Related Work by Others
References to other design documentation, such as the electrical or other contractors' specifications and drawings, that may include work required for, or related to, the AV system installation.

Applicable Standards and Regulatory Requirements
Basic audio, video, networking and control standards that apply to the AV systems installation, as well as any local, state or federal requirements, should be referenced.

Bidder Qualifications
The bidder qualifications are often provided in the Front End section, but this information can also be in the specification section.

Submittal Requirements
General submittal requirements are usually contained in the Front End section. Within the AV specifications, additional information pertaining to AV submittals is included, such as specific requirements for samples, schedule, format, drawing size, electronic formats, architectural elevations and required details for record documents.

Special Conditions or Requirements
There may be a need to address special circumstances that relate to this project specifically for AV. This may include access or security requirements for working in the building, required delivery schedules, acknowledgment of specific site conditions or other requirements.

Chapter 5 — The Design Phase

Part 2 - Equipment

Design drawings for bidding should indicate equipment as generically as possible, except where a design is based on a manufacturer-specific configuration. Design drawings within a design-build process will indicate the equipment more specifically. For a design-bid-build process, the equipment part of the specifications provides more detailed information about each piece of equipment and sometimes for complete subsystems, such as a control system.

There are two basic approaches to writing an AV equipment specification section:

1. The first approach is to describe, in text form, each piece of equipment, including its function and performance requirements. The next step in this approach is to list the acceptable model numbers of equipment that meet the specifications. This method is based on the traditional bricks-and-mortar product specifications. It has been successfully adapted, but can become unwieldy for both the designer and integrator because of the volume of text that is required to be written and coordinated.

   Based on the traditional CSI model, product quantities would not be listed in the specifications. Rather, they would be obtained from the drawings. While this makes sense when specifying gypsum board wall components, it often leads to errors in the bid process if AV equipment quantities are not identified in the specifications.

2. A second approach to writing bid specifications is to provide an equipment list under the product section that identifies at least one manufacturer and model as the basis for design. Any not listed would require approval from the designer if an integrator wishes to provide an alternative.

Part 3 - Execution

The Execution part describes, in text form, how the equipment and system are to be installed, with a focus on the physical installation, overall electronic performance and system testing. Some commissioning requirements and procedures would also be included here. Items typically covered under the Execution part are:

- Quality of Workmanship
- Installation Standards
- Adjustments and Testing
- Testing, Commissioning and Proof of Performance Requirements
- Requirements for Owner and User Training

Creating Design-Build Deliverables

In a design-build solicitation, the design drawings and specification components as described earlier are not included in the documents that are used to contract with the integrator. Instead, there is a package that outlines a scope of work for the integrator with enough information for the integrator to prepare a price for the project. Therefore, the integrator's response to the solicitation will become a major part of the contract documents. By contrast, under the design-bid process the integrator is responding to a package of detailed design information and the integrator's response does not include any significant design information. *Therefore, the design deliverables for a design-build process are mostly produced by the design-build integrator.*

The Design-Build Process

1. The AV Program
On a small project, developing the program may consist of only a meeting or two with the owner and end-users. For a larger process, it may entail a separate contract to produce a full program document similar to the consultant-led option.

2. Design-Build Proposal
The design-build integrator creates a proposal that references or incorporates the program document, which would become the signed contract. In addition to the program document, many integrators provide an equipment list showing equipment prices as well as costs for the installation.

In this case, a Front End section may be part of the contract. Under this scenario, if the integrator contracts directly with the owner, the integrator provides his or her own Front End that covers the administrative parts of the contract.

If, however, a general contractor subcontracts the integrator, then he or she will provide a version of the Front End that they signed as a part of the AV subcontract package. The big difference with the design-bid-build process is that there are typically no three-part specifications involved under this method.

3. Design
Once the design-build proposal is accepted, the integrator must create the AV system design documents for the installers. This process corresponds to the development of shop drawings under the design-bid-build process, with the difference being that there are no contract design drawings from which to work. These shop drawings should contain all the information previously discussed under AV System Drawings.

The integrator is also responsible for the additional base building infrastructure design items that require coordination with the electrical, mechanical and architectural design and installation team. The deliverables for this part of the work are the same as those noted under the Infrastructure Design Process Section.

Clarifying Control System Programming

Because of their design and programming, control systems are potentially complex. They are comprised of two basic types of components: hardware and software. In addition, there are control components that are incorporated into specific audio and video devices that need to be considered during the design process.

To adequately address the control systems design and implementation, the control system may be thought of as five components in the AV systems design:

1. **Control Systems Hardware** — includes the devices produced by a control or AV component manufacturer dedicated to the control system. This includes devices such as control system processors, touch screens, button panels, web servers, Ethernet to RS-232 converters, infrared interfaces, networking equipment and other related control devices. The specification of these should be included in the AV systems design drawings and specifications.
2. **Audio and Video Device Control Options** — includes the options for Ethernet control, RS-232 control or other control access that may require coordination during the selection of the AV devices. These should also be included in the specifications and drawings where needed.
3. **Control Systems Pre-packaged Software** — includes any configuration, monitoring, or operational software that provides the functionality required by the owner and end-users. This category includes "off-the-shelf" software that, while some configuration is required, is pre-programmed and compiled. This may include web server and other software not directly provided by the control system manufacturers. These software components should be specified in the AV design documents.
4. **Control Systems User Interface** — includes the graphical user interfaces for touch screens and web control pages, as well as the design of button panels and other user interfaces to the control system. Under a design-bid-build process, these should be included in the design documents. Good user interfaces are crucial to the success of control systems and even entire projects in many cases; a poor user interface will make the best audio and video system untenable to the users.
5. **Control Systems Custom Software** — includes software that is custom-programmed from scratch. This is the "back-end" programming for the user interfaces that processes the logic behind each button press. It may be advantageous to specify how this software is to be provided in the contract documents.

Figure 20. Matching up Contract Deliverables

Deliverable	Consultant-Led Design-Bid-Build	Integrator Design-Build	Consultant-Led Design-Build	OFE Integrator Install	OFE Owner Install	Description
Program Phase						
Program / Design Contract	O/A	O/A	O/A	-	-	Contract between AV provider and Owner or Architect
Program Information	C	I	C	O	O	
Non-technical System Description	C	I	C/I	O	O	Program Report
Opinion of Probable Costs	C	I	I	O	O	
Design Phase						
MEP, Data/Telecom and other infrastructure design recommendations	C	I	C/I	O	O	AV Infrastructure requirements provided to design team
AV Architectural Plans showing equipment locations	C	I	C/I	O	O	Design or Bid Package for internal use or external bidding
AV System Design Drawings						
Specifications	C	I	C	O	-	
Bid Phase						
Request for Proposal (RFP)	C	-	-	O	-	Request for Proposal; formal document to bidders
Pricing/Quotation	I	I	I	I	-	Accurate pricing provided by the integrator
Installation Phase						
Final Installation Contract	O/GC/C	O/GC/I	O/GC/C/I	O/GC	-	Contract between AV provider and Owner or GC
AV Hardware	I	I	I	O/I/D	O/D	Audio, video and control equipment
AV Shop Drawings, As-Builts & Manuals	I	I	I	I	O	Detailed drawings for installation and final record
Control & Other Software Programming	I/C/O	I/O	I/C/O	I/O	O	Operating code for control, DSP, and other systems
AV Hardware Installation	I	I	I	I	O	Labor to build and install the AV system
Commissioning and Training						
Commissioning	C	I	C	I	O	System optimization and assurance of contract fulfillment
Training	C/I	I	C/I	O/I	O	For both technical and non-technical end-users
Service & Maintenance	C/I	I	C/I	I	O	Ongoing AV system support maintenance and upgrades

Legend:
O   Owner
A   Architect
C   AV Consultant
I   AV Integrator
D   AV Dealer
GC General Contractor

Throughout the design and construction process, there is a host of wide-ranging tasks and deliverables to complete the job. This table provides an overview of which entities are responsible for many of the project deliverables under the various delivery methods discussed.

Including User Interface and Programming in Design Documents

There are several ways that the user interface and programming can be incorporated into the design documents. The system description may be detailed enough to have some control system functionality included, but much more detail is generally needed. The options are:

Option 1: Provide GUI Layouts and Button-by-Button Functionality in Design Documents   BEST PRACTICE

GUI graphical layouts with button-by-button functionality are developed and provided in the drawings and specifications under a design-bid-build process. In a design-build scenario, the GUI and functionality should be developed by the AV designer and provided to the programmer. This practice allows for more comparable bids under a bid option and better GUI designs and functionality in all cases.

Having the GUI for touch screens and web page control of the AV systems designed before the system installation begins (and before it is under contract) with button-by-button descriptions offers several advantages, especially:

- Provides a clear, well-defined and quantifiable scope of work for the control system programmer
- Allows for more "apples-to-apples" comparisons between bids from different integrators in a bid situation
- Allows for the end-users to approve the interface before it is contracted or programmed
- Allows for early end-user training before the systems installation is complete

Option 2: Provide GUI Layouts Only in Design Documents   ACCEPTABLE APPROACH

Graphical User Interface (GUI) graphical layouts are defined in the drawings and specifications in the design-bid-build process, or developed by the AV designer and provided to the programmer under a design-build process, but they appear without button-by-button descriptions. While this may be better than providing no graphical layout, the button-by-button descriptions are crucial to getting the functionality that is required by the users.

Option 3: No GUI Layouts Provided   NOT RECOMMENDED

General functionality is defined in the specifications, but the layout and operational details are left to the programmer based on the audio and video block diagrams and the overall system description. This can result in poor user interface design and misunderstandings of the required functionality.

Budget Management

During infrastructure design, the responsibility for budget management normally rests with the associated base building design engineer and/or the contractor/construction manager who may be involved. However, during the design phase, it is the responsibility of the AV designer to provide a design that meets both the functional requirements and the budget established for the systems.

Depending on how the original opinion of probable cost was created, the AV designer should be working with a budget that is consistent with the functionality and quality that the users expect. In a smaller, short-term design-build scenario, the system budget may have been developed during the program phase and based on the proposed equipment list, plus anticipated

installation costs. This scenario makes it relatively easy for the design to be consistent with the budget. On the other hand, with a longer-term design-bid-build process, the budget is generally not based on a detailed design or equipment list, so anticipated system costs need to be monitored as the design develops.

Cost monitoring

Because of variations in project size and processes used by different AV firms, using a uniform methodology in monitoring costs is not only impractical, but also unworkable.

Some AV firms have systems in place to facilitate the process. These internal procedures[23] enable an AV project manager and the designer to check the anticipated system cost against the working budget at appropriate points in the design phase — but even these systems have their limitations. For a firm working from previously developed design templates, it is relatively easy to check the budget. For firms without an extensive template database, or when working on customized or unusual projects, it may not be possible to make a first pass at a budget check until a significant portion of the design work has been completed.

In general, there should be at least two budget checks during the design phase: first, when there is enough information to create a fairly detailed equipment list, and second, when the design documents are nearly or fully completed.

Budget checks should also be performed before each review. Reviews may be conducted by the owner, program manager, end-users or the AV consultant (as in the case of a consultant-led design-build process). These checks can be handled on a piecemeal basis (for larger projects where parts of the systems can be separated and reviewed as they are completed), or they can be conducted for the entire system.

Ultimately, the goal of budget management is to create a package that will require little or no changes (barring any unforeseen events) in meeting the users' functional requirements.

Step Four: Making the Most of Review

Reviews provide information, enable process checks, and serve to garner support from the individuals who will ultimately use the systems. Conducted properly and openly, reviews go a long way toward helping to build a successful system.

AV Designer — Internal Reviews

A best practice among AV designers is to have a different person in the AV firm take responsibility for the AV systems design review. It is also a good idea to do the budget check at the same time the internal review is being conducted. Dovetailing this effort enables many design errors to be caught prior to bidding or quoting a project.

> **Guidelines for design: the Dashboard for Controls**
>
> The Dashboard for Controls project is an ongoing project spearheaded by InfoComm's Technology Managers/End-Users Council and developed in collaboration with other councils, including the Independent Programmers Council and the Independent Consultants in Audiovisual Technology (ICAT) Council. Its goal is to establish recommended guidelines for design of controls that would make their operation by end-users as easy and intuitive as driving a car (hence the analogy to a dashboard). A white paper and templates have been created as of publication of this book, and they are available for download at www.infocomm.org/dashboard. As of this writing, a Design Guide and a Design Reference document are available that relate to mid-sized touchscreen interfaces.

---

[23] These often take the form of templates incorporated into a documented process. For any budget checking to occur, there must be enough design information in place. Sufficient detail concerning the type and cost of the significant pieces of equipment is needed in order to create a cost estimate that is more detailed and accurate than that provided during the program phase.

Best Practice

Get organized for design reviews

How the owner/user is organized, and/or how the project is contracted, will have an effect on how the owner review process takes place. For owners and end-users with non-technical personnel, the review may require face-to-face meetings to explain the documents and interpret different parts of the system in layman's terms. In other cases, the reviewers may be more technically capable and can review the documents without significant assistance. In some cases (particularly on larger projects), a meeting is required at the conclusion of each review to discuss any comments or issues that have arisen.

Owner/Project Team Reviews

In the design process, it is important that the owner participate in the process as much as possible. This achieves the following:

- Provides the owner and end-users with an understanding of the systems to be installed. This may apply more to the owner's technical staff than to the end-users, but the end-user organization must be prepared for the operation of the new systems.
- Builds a sense of buy-in on the design decisions that are made during this phase, particularly user interface elements and equipment performance (such as projector resolution and quality). This is an important part of managing the expectations of the end-users and keeping the project stakeholders on the same page.
- Obtains feedback that may be important to the scope, either where the program was in error or the needs of the users have changed. This feedback may suggest mid-course design changes that will be easier to address at this stage, rather than later when significant changes might be required. This aspect of the review is a key quality control measure.

Review Participants

A number of project participants are typically interested in reviewing the AV design documents. Each has different interests in the review:

Architect

By contract, the architect is the conduit for AV design reviews who will pass information on to others in the project team. In addition, the architect has his or her own interest in reviewing the drawings and specifications primarily related to base building coordination issues, such as display locations, loudspeaker layouts and equipment mounting. Architects may also need to see the completeness of the package from the standpoint of verifying the completion of the design contract terms.

Construction/Program Manager

As a stakeholder in the outcome of the overall project, the construction or program manager is interested in reviewing the AV system design documents both from an administrative standpoint and from a technical coordination standpoint.

Owner Administration

The owner's administrative representatives (executives, accountants or contract specialists) may also want to review the AV design — either in light of contract terms or because of a personal interest in the project and its progress.

End-User

End-users may review the design documents if they have an understanding or interest in them. The end-user's technical staff will want to review the technical aspects of the design, including system configurations, equipment selections and any support-related systems or networks. The ultimate end-users of the equipment, such as instructors or presenters, who are often non-technical, may not be able to comment on the more technical aspects of the system design, but may want to respond to the system functional descriptions, equipment locations and any user interface designs.

Other Reviews

Other stakeholders may have an interest in reviewing the documents. Examples might be a funding organization, the building committee, or other interested party not normally involved in design reviews.

In some cases, there may be a requirement that an AV industry peer of the owner's, architect's or construction manager's choosing review the design documents. This is a constructive process in which one qualified professional reviews another qualified professional's work. Errors can be identified and suggestions for valid design approaches may surface.

Whoever is requiring the peer review must be careful to select a peer organization that is properly qualified for the review and to ensure that there is no competitive pressure between the two organizations. It is not recommended that the reviewing AV firm be one that was considered but not selected for the project, or that it be a direct competitor in the same geographic area as the designer.

Moving on to Construction

Once the design is complete and review comments have been addressed, the system is ready for construction. Under a design-build option, this means moving right into the purchasing and assembly of the systems and beginning the on-site work. For those involved in design-bid-build, there is one more step, the bid process, before moving into the Construction Phase.

Chapter 5 Checklists

DESIGN PHASE INFRASTRUCTURE COORDINATION CHECKLIST

During the design phase, there are a number of coordination issues to be addressed between the AV designer and the base building design team as noted in Chapter 5: Design Phase. The major items are listed in this checklist.

Architectural Construction, Space Planning and Interiors

- ❏ Develop seating layouts and room orientations
- ❏ Develop furniture layouts
- ❏ Provide display types, sizes and locations
- ❏ Recommend room finishes
- ❏ Perform sight line studies
- ❏ Provide acoustics and noise control recommendations
- ❏ Develop AV-related millwork details
- ❏ Provide assistance with anticipated AV-related infrastructure costs

Provide architectural integration details for:

- ❏ Projection screen installation
- ❏ Loudspeaker mounting
- ❏ Structural mounts for displays and other devices including loads
- ❏ AV integration into furniture
- ❏ Wall-mounted racks, monitors and other devices

HVAC

- ❏ Provide head loads from AV equipment
- ❏ Coordinate duct runs and terminal device locations
- ❏ Provide noise criteria for AV spaces

Electrical, Lighting and Data/Telecom

- ❏ Convey conduit and backbox type, quantity and locations
- ❏ Provide power requirements and power grounding details
- ❏ Provide lighting layouts and/or criteria for AV spaces
- ❏ Provide data/telecom requirements for AV

Administrative Coordination

- ❏ Develop anticipated AV system costs
- ❏ Provide AV design and installation schedule

Administrative "front end" documents
A. Invitation to Bid
B. Request for Qualifications (RFQ)
C. Bid Response Form
D. General Conditions
   1. Project work conditions and terms
   2. Definitions of terminology
   3. Materials and workmanship
   4. Safety and accident prevention
   5. Permits, regulations and taxes
   6. Insurance
   7. Bond
   8. Overall quality control
   9. Submittals
   10. Substitutions schedule
   11. Changes in the scope of work
   12. Claims and corrections pertaining to the work
   13. Penalty clauses.
   14. Warranty
   15. Non-discrimination and affirmative action
   16. Contract termination options
   17. Arbitration
   18. Invoicing and payments
      - Terms and payments
      - Deposits and mobilization fees
      - Communications procedures
      - Progress payments.
      - Final payment
   19. Project Closeout
   20. Record Documents

Architectural and Infrastructure Drawings
A. Architectural Plans and Details
   1. Reference floor plans
   2. Reference ceiling plans
   3. Sections of major spaces
   4. AV device location and coordination floor and ceiling plans showing room-mounted device locations, rack layouts, seating layouts and critical dimensions.
   5. Millwork and furniture integration details for AV equipment
   6. Structural accommodations for AV equipment
B. AV Electrical and Communications Floor, Wall and Ceiling Plans
   1. AV-related power outlets and junction boxes
   2. AV signal conduit risers and plans
   3. AV signal junction box, floor box and backbox location plans
   4. Voice/data, CATV, other related existing outlet locations

---

OUTLINE OF TYPICAL RFP / DESIGN PACKAGE COMPONENTS

This outline represents a comprehensive example of a complete AV design package for full design and bidding. Some components may be used for a design-build process minus the AV system drawings and specifications. Some projects may require additional items.

OUTLINE OF TYPICAL RFP / DESIGN PACKAGE COMPONENTS

— Continued

AV Systems Drawings

1. Title and index
2. Legend of symbols
3. Typical power, grounding and signal wiring details
4. Floor and reflected ceiling plans showing device locations
5. System functional diagrams
6. Rack elevations
7. Custom plate and panel details
8. Miscellaneous details and elevations including
    - Large scale plans, such as equipment or control room plans
    - Architectural elevations showing AV devices, their location and their relationship to other items on the walls
    - Custom-enclosure or mounting details for projectors, microphones, loudspeakers, media players and others
    - Furniture integration details
    - Speaker aiming information
    - Any special circumstances or details that may be required for the installers to properly understand the design intent

AV System Specifications

A. System Description including excerpts from AV program report
B. Control Systems Graphical User Interface (GUI) designs with detailed functional and/or button-by-button description
C. Three-part CSI Specifications
   1. Part 1 — General
      - Project Description
      - Work Included
      - Alternates
      - Related Work by Others
      - Applicable Standards and Regulatory Requirements
      - Bidder Qualifications
      - Submittal Requirements
      - Special Conditions or Requirements
   2. Part 2 - Equipment
   3. Part 3 - Execution
      - Quality of Workmanship
      - Installation Standards
      - Adjustments and Testing
      - Testing, Commissioning and Proof of Performance Requirements
      - Requirements for Owner and User Training

CHAPTER 6
The Construction Phase

Step One: The Construction Kick-Off Meeting
Step Two: Preparing for Submittals
Step Three: Procuring the AV Equipment
Step Four: Preparing the Site
Step Five: Pre-Assembling and Testing the AV System
Step Six: Site Installation
Step Seven: Finalizing the Documentation

CHAPTER SIX

The construction phase is focused on three key processes: coordination, procurement and installation. Finally, the prepared designs are translated into physical form, and the systems are brought into functionality. During this phase, there are tasks associated with both the base building construction, which includes the AV infrastructure, and the AV systems electronics procurement and assembly.

Coordination, scheduling and sequencing that are key to the entire construction phase are reviewed in this chapter, along with typical project communications and how they relate to changes in the AV installation contract.

The construction phase is described in relation to the AV systems, following the steps that the integrator takes from contract award to a complete installation ready for commissioning. The steps are:

1. The construction kick-off meeting
2. Preparing for submittals
3. Procuring the AV equipment
4. Preparing the site
5. Pre-assembling and testing the AV system
6. Site installation
7. Finalizing the documentation

Most of the AV integrator's work during this phase is related to the electronics rather than to the base building construction. Shown below is the delineation between work of the general contractor and the non-AV subcontractors at the construction site and the work of the AV integrator during this phase.

Construction Phase Activities

Base Building Construction
General contractor and other trade subcontractors

- Base building construction including the walls, floors, ceilings and structure
- Space demolition and renovation activities
- Installation of architecturally integrated AV items, such as projection screens, electronic marker boards and loudspeaker supports
- Mechanical, electrical, plumbing and other major systems installations including conduit and power, lighting and HVAC noise control measures
- Data/telecom, security, fire alarm and other low voltage systems installation including data and other outlets required for the AV systems.
- Furniture, millwork and casework purchase, fabrication and installation including items that may house or support AV electronics

AV Systems Electronics Installation
AV integrator

- On-site AV cable installation and termination
- Installation and pre-testing of AV electronics and wiring into equipment racks either on-site, in the AV shop, or both
- Installation of AV electronics and wiring into furniture, millwork and casework including DVD players, switchers, video interfaces, control systems devices, connectors and other electronics
- Installation of peripheral AV electronics into spaces including projectors, cameras, displays, loudspeakers and microphones
- Wiring and testing of all on-site equipment including racks, connector plates and peripheral equipment

These two tracks of tasks occur concurrently, culminating in the finished building spaces with the AV systems installed.

Project Coordination, Scheduling and Sequencing

Just as coordination is required between the consultants during the design phases, a considerable amount of coordination is also needed during construction to avoid or resolve potential conflicts in the design documents or any that may arise due to conditions in the field. Depending on how the project is contracted, the GC and/or the construction manager are responsible for coordinating the trade contractors during their work on-site. The architect, owner and various trade consultants are also involved to help resolve issues as they come up. As in the design phase, conflicts that are caught and resolved prior to starting work are certainly the easiest to resolve.

To facilitate coordination, a general project schedule is typically established during the design and bid phases and is detailed and finalized by the owner's project manager as work begins. Each trade or discipline must have a corresponding schedule of work (developed by each team's project manager), and this information is inserted into the main schedule.

As with the other trades, AV must be integrated into the overall schedule. The AV integrator is usually involved during the middle and final portions of the construction schedule when AV cable pulls, structural rough-in, and other AV-related infrastructure are installed.

The duration of these tasks is just as important as their sequencing throughout the project. For instance, conduit and back boxes for AV cabling must be in place before cabling can be put in place, but the cable should be pulled before installing the ceiling grid where cabling runs overhead in rooms and corridors.

What makes AV installation different from most other trades is that the bulk of the audiovisual system (i.e., the electronics) must be installed after all other trades are complete and the site is clean, because the equipment is so sensitive to dust, moisture and temperature.

Coordinating AV Infrastructure Makes it all Work

Because of the impact that AV has on so many other trades, coordination is required with all other contractors on the job. Much of the responsibility for coordinating AV-related issues falls upon the AV providers. A successful AV project depends upon the ability and expertise of the AV professional and the other construction team members to dovetail the AV work with the other building trades. Experienced AV providers know that the end product will benefit from close coordination with the special trades and services.

AV project coordination during this phase covers many different aspects of the project. Examples of areas that need particularly close coordination include:

- Millwork that integrates with AV systems must be built to spec and tolerance.
- Conduit, raceways, cable trays, cable paths and electrical power must be located and sized correctly. The schedule of this work is also critical to the AV integrator.
- Lighting fixture placement and illumination properties should be checked for impact on AV systems, particularly projected images.
- Lighting dimming systems and any interface(s) must be confirmed.
- Installation of HVAC ductwork and terminal devices must be monitored for conflicts with the AV installation
- Dimensions of loading docks, elevators, doorways, corridors and passageways must be large enough to allow delivery of equipment, particularly large rear projection screens and millwork that is fabricated off-site.

The project manager's tool: the Gantt chart

The most common way to create a project schedule is by using a "Gantt chart," Invented by Henry L. Gantt in 1917, this chart includes horizontal bars, showing tasks and their relationship on a timeline. A Gantt chart is organized in logical groupings of tasks in a format similar to an outline. In traditional project management terms, this is referred to as a Work Breakdown Structure (WBS). A WBS is a hierarchical notation that defines the tasks, activities, deliverables, and milestones required to complete a project. Each entry in the WBS has properties including, but not limited to, dependencies, budget, time duration, resource requirements, assignments and schedule restraints.

Of particular note in the construction schedule is the AV trade. AV cable pulls, structural rough-in, and other infrastructure should be accomplished during construction when possible and practical. Other than infrastructure, the bulk of an audiovisual system must be installed after all other trades are complete because of its sensitive nature. See appendix for an example of a sample AV system project schedule.

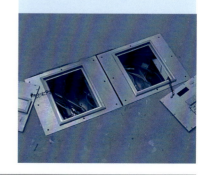

BEST PRACTICE

AV consultant and integrator work as a team

Open communication and respect between the AV consultant and integrator are essential under any method involving both, especially within the design-bid-build model. To deliver a technically correct product while maintaining the schedule and budget, these professionals must work together as a team.

The consultant brings a broad knowledge of AV solutions and an intimate understanding of the client's needs. After evaluating those needs, the consultant applies and defines AV solutions for construction.

The integrator brings a depth of field experience, specific product knowledge and a trained technical staff, and is responsible for delivering a complete system, based on the design specifications from the consultant.

As professionals, the consultant and integrator are committed to communicating any concerns with the AV component of the project to each other as well as to the CM or PM and those to whom they are contracted. The sharing of ideas, technology, product demonstrations, solutions and samples promotes teamwork, opens the way for better products, ensures greater client satisfaction and ultimately promotes the AV industry as a whole.

(Areas of Coordination, cont.)

- Existing facility use must be considered, including any spaces that might be affected by work on the project. For instance, drilling or cutting may create noise on other floors or equipment deliveries may only be allowed during off-hours.
- Ceiling layout and material, location of lights and sprinkler heads must be coordinated with installation of ceiling-mounted AV systems such as loudspeakers, projectors, cameras and projection screens.
- The scheduling of walls and ceiling installations must be monitored for sequencing of the installation of wiring, structural mounts and AV devices.
- Union and trade organization agreements must be honored.
- Acoustic treatments, which can significantly affect the performance of audio systems, must be considered. Wall, floor, ceiling, window, furniture/upholstery and other treatments all affect the acoustic behavior of a space.
- Other work and treatments that may create a hazardous environment for AV equipment must be complete before equipment is delivered to the site. This work includes carpentry, sanding, painting, concrete work and ceiling installation in areas where AV equipment is to be stored or installed.
- Data/telecom services must be installed and activated before functional testing of AV systems.

These coordination issues may be the responsibility of the AV consultant, the AV integrator, or both, depending on which delivery method is being used for the AV systems.

The Interdependence between AV and Data/Telecom

Early on, the AV integrator forges a strong working relationship with the local data/telecom department and the data/telecom contractor, if there is one on the job. The owner's role in this relationship is important because the owner's staff typically provides and configures the network electronics that will form the data network.

Although data/telecom requirements (usually in the form of cabling only) are normally specified during the design phase, it is the responsibility of the AV integrator to ensure that the needs specific to the installed systems and equipment are met. Both the physical wiring and the electronic network configurations must be coordinated in relation to AV. For example, connectivity via Internet Protocol (IP) may require other services, such as routing, Quality of Service (QoS), priority and other network-related management in order to provide acceptable delivery of Internet video.

Data/telecom services are required for transport of many signals, including video, audio, conferencing and control. The following represent data/telecom services and physical wiring that are used for transport of AV needs but are typically provided by the owner:

- Ethernet connectivity, used for communications by control systems, video-conferencing, projectors, and many other AV devices and systems, including IP addressing and subnet schemes
- Network segmenting, routing and VLAN/VPN schemes
- Communications services from data and telephone service providers, including ISDN, BRI/PRI, E1/T1, and analog telephone, primarily for conferencing use
- Cable TV service activation
- Structured cabling for AV over either unshielded twisted pair (UTP) copper or fiber-optic cabling

Scheduling of the installation and activation of these items is important to the AV services being installed.

Owner and End-User Responsibilities

In addition to coordination and scheduling with the installation team, there are important AV-related responsibilities that involve the owner during the construction phase. These range from administrative issues to provision of services. A few of the activities in which the owner's team should be involved are:

- Timely processing of bid and contract documents
- Prompt payment of invoices within contractual obligations
- Provision of timely access to the site
- Provision of external electric and HVAC utility services
- Activation of data/telecom services
- Provision and activation of cable/satelite TV services (as required)
- Delivery of any owner-furnished equipment and furniture
- End-user move-in coordination

Attention to these issues will ease the construction process for the installation team as well as for the owner and end-users.

Coordinate the move-in schedule with the AV installation

The AV equipment requires a clean environment and the system cannot be fully commissioned until the facility construction is completed. This may create an overlap of the AV installation on-site and the move-in by the facility's end-users. The "first use" and other scheduling of AV events by users should be closely coordinated and sequenced to allow for completion of the AV systems installation process.

Project Communications and Contract Changes

Consistent and accurate communications are crucial to a successful project. Many of these are related to or become a part of the various contracts between the contractors, consultants and the owner. There are forms and formats from several industry-standard sources (see Appendix for a list of available AIA forms), as well as those developed by companies and contractors for their own use. The acceptable forms should be defined in the RFP or contract and agreed upon prior to the start of the project.

All communications, reports, transmittals and other documents should include project-specific references, including the owner's name/address, specific installation location(s), contractor name(s), project numbers and other appropriate contact or contract information. Creating a set of standard approved forms that are project specific and pre-addressed at the beginning of the project saves time and speeds up the process.

The following are descriptions of some of the more important forms that are typically required for use by the AV and other contractors and consultants in a construction project. At times, the names of the forms vary from those used here, but the purpose behind them remains the same.

Letter of Transmittal

This form is used whenever documents, drawings, samples or submittals are sent. It clearly indicates the addressee, sender, contact information, a list of what is being sent (including date or revision number) and any expected action to be taken by the receiving party. This form is used whether the items are sent by mail, courier, overnight carrier or fax.

Requests for Information (RFI)

As the project progresses, questions inevitably arise about the project. They generally revolve around three basic types of issues: A design issue, a site issue or an owner change or request. The structure normally set in place for this process is the Request for Information, or RFI.
This process is usually based on a paper or electronic form established for the project that includes who the RFI is from, who it is directed to, and a space to enter the question and the response.

Some RFIs are simply resolved by a clarification from the recipient of the RFI without a change in anyone's contract. Others may need resolution through a change in the construction contract. In the latter case, other structured communications such as a Change Order may be generated.

Request for Change (RFC)

An RFC is submitted (ultimately to be approved by the owner) if the integrator or consultant wants to change contractual obligations, equipment models or specifications, or system design. When an RFC is generated (or answered) by the integrator, pricing and impact throughout the project must be included.

Issues that can trigger an RFC
- Change in intended use of the system
- Discontinued product
- Architectural, mechanical or millwork changes
- Discovery of system or product incompatibilities or function

The RFC can be submitted by any member of the project team, although in an AV project, the integrator or consultant most commonly creates it. An approved RFC then becomes a Change Order.

Change Orders (CO)

Despite extensive due diligence during design and bidding to ensure an appropriate system design, there may be a need for design and contract changes, requested with a Change Order, as the project unfolds. Because of its ability to change the contract scope and pricing, this is arguably the most important form used during the construction phase. Here are a few of the many reasons for AV system change orders:
- Changes or clarifications in anticipated use by end-user personnel
- Architectural, millwork, finish or other physical changes to the installation site
- Design conflicts, omissions or errors
- Change in product availability or specifications
- Availability of new products or technologies
- Discovery of hidden site conditions
- Budget adjustments
- Schedule changes and delays by others

The change order procedure, and those authorized to approve them, are established by contract at the beginning of the project. Since changes can dramatically alter the project schedule or budget, the CO's should be processed in a timely fashion.

Change orders almost always affect the overall value of the project. They can have an impact on financing, leasing, insurance and bonding costs, so guarantors should be notified if a change order causes the project value to exceed the original budget. CO pricing can be included in the Request for Change or handled on a "time and materials" basis (pricing based on the actual labor and materials required for the work).

Construction Change Directive (CCD)
A CCD (sometimes also called a Field Order) is usually issued as a result of time-critical conditions or events in the field, to trigger immediate work related to a change needed in the contract. This form is typically used by the owner or the GC to bypass the more formal RFI-RFC-CO process that is normally required for a contract change.

Step One: The Construction Kick-Off Meeting

After the project is awarded, but before the start of work, a kick-off meeting should be held to get the construction phase team members acquainted, discuss the project and establish how project communications should occur.

Depending on how the contracts are structured, the owner's project manager, the construction manager or the general contractor may preside over this meeting. The stakeholders of any contributing team that affects the outcome of the project should be in attendance, but at a minimum, the architect, general contractor, electrical contractor, an owner representative and the AV providers should be present.

During the kick-off meeting, logistical protocols and communication channels should be established. These include proper forms, authorizations (financial and design), process for payment, RFI handling and proper lines of communication. Unless clarified or revised along the way, these protocols must be consistently maintained throughout the life of the project.

Other issues that should be addressed during the kick-off meeting are:
- Coordination and sequencing issues that are specific to the AV integrator's work, as described earlier in this chapter
- Any owner concerns or project politics that may need to be considered during the construction phase
- Contract issues needing review, such as contractor billing procedures, warehousing and bonding requirements and project schedules (particularly as they relate to any penalty clauses)
- AV equipment delivery and installation logistics including OFE delivery, site cleanliness requirements and equipment security concerns
- Scheduling of the "clean and clear dates" for AV equipment installation in finish spaces (final power and AV running)

Handling changes in the AV system

Changes to the AV system contract may involve electronics, wiring, system configurations, furnishings or simply project schedules. It is important for the AV integrator to understand that changes or substitutions which have not been approved by the proper authority on the project are purchased at the AV integrator's risk. Approvals are usually required by the GC, the owner and/or AV consultant for any AV-related contract changes.

If a change includes a substitution for a product already purchased and installed, this could appropriately involve a restocking fee and labor charges, not to mention an impact on the schedule. While change orders are notorious for being costly, many changes can include deletions that would create credits to the contract. Pricing for any change should be reviewed carefully for scope and accuracy before being approved.

Keeping Up After the Kick-Off Meeting

Coordination with other trades is essential to the success of a project. It is imperative that all trades, designers, contractors, sub-contractors and others remain in constant contact. The coordination is typically accomplished through weekly construction/coordination meetings so all disciplines can accommodate one another's work and schedule.

During weekly project meetings, each team's project manager provides a status report of that team's progress and any issues that may have an impact on other teams. Information is exchanged, schedules are updated and coordination issues are resolved. The AV professionals may be called upon to review trade shop drawing submittals for coordination purposes. The AV providers may not need to be at every weekly meeting, particularly at the beginning of construction when they have little or no involvement on-site. They should start to visit regularly as the construction of their spaces and rooms are built to check for field changes or irregularities that might affect their work. Their work begins in earnest with the cable installation, at which point the AV integrator brings equipment and personnel on-site for system installation.

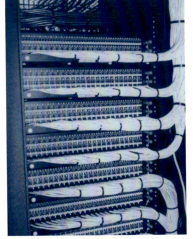

The weekly meeting organizer keeps minutes of each project meeting, recording all issues discussed at the meeting, any conflict resolutions and action items that may be required by different members of the project team. This record should be distributed to all participants as soon as possible after the meeting. Each trade team leader may also maintain a record of meeting proceedings as they pertain to the leader's own team for their internal use.

Step Two: Preparing for Submittals

For both the design-build and design-bid-build methods, documents and samples almost always need to be prepared and submitted to the owner or the design team for review and approval. Generically, these are called submittals in construction contracts. For the AV systems, submittals include several typical components:

- Shop drawings indicating the fully detailed AV system configuration from which the installers will build the system. These should indicate items and locations of required infrastructure to be used by the integrator but provided by others (e.g., conduits, blocking, hangers, power outlets, etc.)
- Equipment product information that describes the equipment being used on the project
- Samples of finishes for items that may be visible to the end-users and require coordination with other room finishes and furnishings
- Samples of technical work, such as terminated connectors or cabling that illustrate the contractor's understanding of the detailed design intent

The items listed above are often required by contract to be approved before procuring and installing any equipment. For a faster project process, there may be a series of submittals to allow for certain items to be ordered soon after the contract is awarded.

Audiovisual Best Practices

There may be other submittals required by the contract later in the process, including:
- Final as-built record documents that consist of the shop drawings updated to reflect any deviations, adaptations or substitutions that may have occurred during construction
- A binder set that includes all user manuals provided with the AV electronics used in the system
- Additional custom user procedures indicating common setup and operating procedures for the system as installed

Preparing Shop Drawings

During the design phase, AV specifications, sketches and drawings are generated. These are typically detailed in a manner sufficient to show design intent, but they are not suitable for final fabrication and installation. From this stage, shop drawings need to be developed.

If an AV consultant participates in the project, he or she typically generates one-line signal flow drawings with a set of specifications. It may be helpful to provide these drawings electronically to the AV integrator with a use-limitation or license agreement prior to release. This license allows the integrator use of the drawings as reference for the specific project, while protecting the

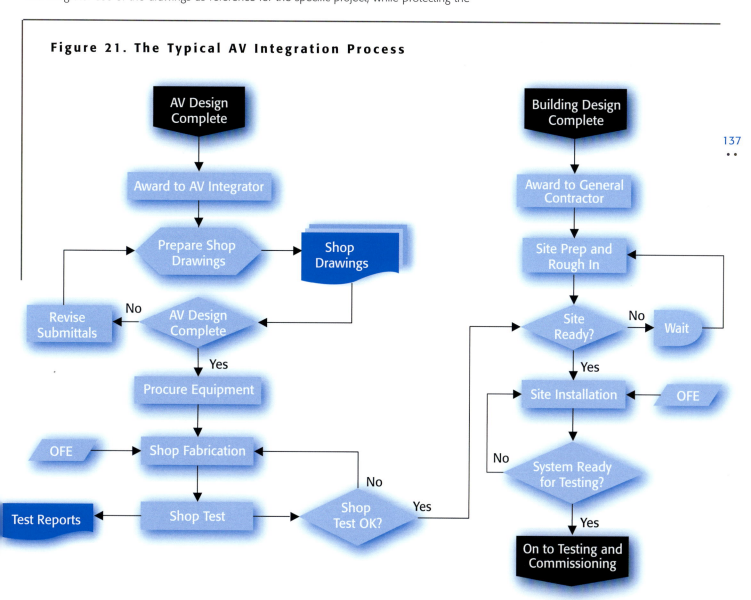

Figure 21. The Typical AV Integration Process

consultant's intellectual property, i.e., the design.

Documents that the consultant may furnish include audio, video and control signal flows; architectural plans, elevations and details; graphical user interface (GUI) designs; and digital signal processing (DSP) software configurations. These documents are used as a reference and a basis for development of the shop drawings. Electronic versions of architectural background plans and MEP drawings should be available from the architect and other design team members.

What to Look for in Shop Drawings

AV system shop drawings are a fully developed, complete set of drawings, which contains all technical aspects and details of the system to be built. Besides the fact that shop drawings — and subsequent as-built drawings — are normally required by contract, it is advisable to generate shop drawings for all projects.

Drawings are normally prepared electronically in an appropriate CAD or drawing software package. Add-on packages are also available that create AV-specific drawing environments.

A complete set of drawings consists of the following:
- Cover sheet, showing project references and listing of included drawings
- Dimensioned/scaled architectural plan, reflected ceiling, elevation and detail drawings as applicable, showing specific location and details of all AV components and systems to be installed on-site; items by others which require coordination should also be shown
- Reflected ceiling plans of all spaces, showing all AV devices and all items by others which require coordination with AV devices
- One line drawing(s) of the system
- Detailed signal flow drawings for audio, video, control and AV power, including:
  - All equipment, showing model numbers and all inputs/outputs
  - All wires and conductors with wire designations
  - Terminal blocks, connectors/types and color codes
  - Input/output/port assignments, matching manufacturers' designations
  - Global details for connectors, pin-outs, patchbay wiring, and grounding
  - Details for any custom circuitry
- Equipment rack details, including elevations, ventilation/cooling, mounting, terminal block locations, AC and signal wiring paths
- Patchbay elevations, showing labeling of all jacks
- Structural drawings relative to AV systems, showing all components, details, dimensions, weights/loads and attachments
- Dimensioned mounting details of all optical and display systems (projectors or screens)
- Detailed drawings of all custom-fabricated plates, panels, devices, labels, patch cables and other custom items

The Simplified One-Line Drawing

By nature, shop and as-built drawings are burdened with details, often making them far too complex for basic understanding and troubleshooting of an AV system. If this is the case, a one-line drawing should be created for the system that is provided in addition to the detailed as-built drawings based on the shop drawings.

A one-line drawing is a simplified, easy-to-read representation of the system, indicating the signal paths from all sources to all destinations or displays. This drawing serves as a navigational aid to understanding or troubleshooting the system. If the system is too complex to be represented in its entirety on one sheet, logical sub-systems should be defined (e.g., video/audio/control or room 1/2/3) and documented accordingly. A copy of the one-line drawing should be posted near the primary system rack.

As-Built Drawings and Manuals
A set of as-built drawings should be prepared after nominal site testing has been completed. This drawing set should include all changes (especially those occurring during shop fabrication/test and site installation/test) that were made to the shop drawings.
In addition, the user manuals provided with the individual AV components arranged in alphabetical order by manufacturer should be compiled into a set of binders. These are made available for use by the technical staff for troubleshooting, repair and special equipment setups. The manuals should be augmented by a custom operation guide that is provided to the end-users for basic setup and operations in the rooms served by the AV systems.

Custom Operation Guide
It is appropriate to create a system-specific custom operation guide for all but the simplest of systems. This guide is mostly for the end-users but also serves the technical staff for both setup and operation of the system. It should include:

For the end-users:
- "Flash-card" style instructions for all basic system operations (e.g., playing a DVD, connecting a computer for display, volume and other audio settings, etc.)
- Basic troubleshooting guide in case of system malfunction, including common user errors and equipment failures, as well as how end-users can get help

For the technicians:
- List of consumable spare parts (projector lamps, air filters, etc.)
- Description of recommended service needs and intervals
- Warranty statement, including:
    - System warranty start date, conditions, and terms
    - Summary of manufacturers' warranty coverage
    - Description of extended warranties and service plans as purchased with the system

Getting Submittals Approved
Shop drawings should be submitted to, checked and/or approved by the consultant, architect, owner, millwork shop and any other project participant as applicable or required by contract documents. The approval process, determined and documented at the beginning of the project, must be followed. Each project may have a different approval structure. If an AV designer has generated a specification, design drawings or program document, he or she should be involved in the approval process.

If possible, the AV integrator should forward an advance or informational copy of all submittals to the consultant in consultant-led projects. It is also a best practice for the AV integrator's team to review and/or approve all shop drawings internally prior to submittal to the project manager and other trades so all contributors have input with respect to their specific tasks. These internal contributors include those responsible for rack fabrication, test/alignment, site supervision, software programming and quality assurance.

Without appropriate approvals, the AV integrator is at risk for conflicts between the intended design and the AV system.

Figure 22. Relative Schedules During Construction

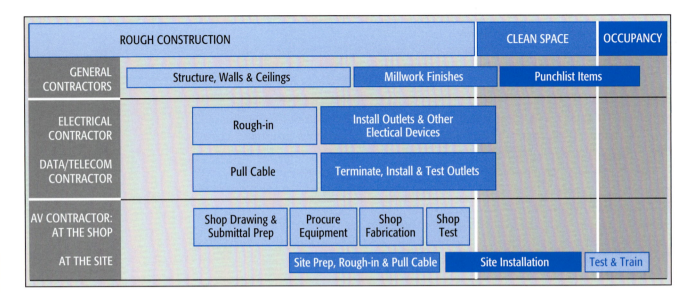

This chart shows the sequence of the AV integrator's work in relation to work by the general contractor, electrical contractor and the data/telecom contractor during the construction phase. In particular, note the need for clean space at the end of the base building construction process and the potential overlap into the end-users' occupancy of the facility.

Step Three: Procuring the AV Equipment

A by-product of the shop drawing process is the assurance that all equipment meets the functional and connectivity requirements of the system. For that reason, best practice, if not the contract, dictates that all goods are ordered only after shop drawings are approved. This means that the submittal approval process should be completed within a timeframe that allows equipment and materials to be ordered so that they arrive in time for shop fabrication and site installation.

In some cases, however, a short schedule may require that long-lead items be ordered prior to approval of shop drawings. If this is the case, pre-ordered goods need to be sufficiently researched to assure that they will meet the design requirement. Certain goods can be ordered in advance by the integrator, and if necessary, by the owner prior to award. These include any items with a lead time, such as custom-built and specialty items and millwork. Any ordering of long-lead items by the integrator should not be done without written approval from the owner and/or the design team, regardless of whether or not this is expressly required in the integrators contract.

For all equipment to be furnished in the project, the integrator should have a direct relationship with the manufacturer, i.e., authorized dealer and/or service center. The integrator's technical team should be appropriately trained for the proper implementation and service of all items prior to inclusion in a project. When specific models or specialty items to be used are not within the integrator's authorized product offering, arrangements should be made in advance for the proper supply, setup and service.

Accommodating Owner Furnished Equipment (OFE)

In some instances, the owner may furnish equipment for the AV project. This equipment could be a combination of any of the following:

- Newly purchased
- Equipment that is currently installed locally or elsewhere
- Specialty furniture or fixtures, purchased from others, to be integrated into the finished system
- Warehoused items, spares and other equipment previously purchased but not yet in use

During the design process, these items should be identified and the status of each item stated in the proposal. After award, the integrator dismantles and/or takes delivery of all OFE and evaluates each item. Evaluation should include physical inspection, testing of all functions and any appropriate cleaning or refurbishment. The results of the evaluation should be compared to the original manufacturer's specifications. A report citing manufacturer, model, serial number and status of all items should be submitted to the owner.

If evaluation of OFE deems it unsuitable for use in the intended system, any repair or replacement required is normally considered outside of the project scope. Resolution of this is negotiated between the AV integrator and the owner.

Step Four: Preparing the Site

While the other trades are constructing the spaces and systems that will serve as the AV infrastructure, the AV contractor requires that certain activities be performed to prepare the site to receive the AV systems. This involves monitoring the installation of electrical and mechanical systems as well as initial cable installation and providing and/or installing architecturally integrated items such as projection screens and ceiling mounted loudspeaker assemblies.

In some design-build cases, the integrator may be required to install some of the infrastructure under his own contract. In this case, the AV integrator may subcontract to an electrical or data/telecom contractor to install conduit, backboxes, power outlets and data jacks that may be required. Mechanical and architectural infrastructure installation, however, is normally installed under separate contracts.

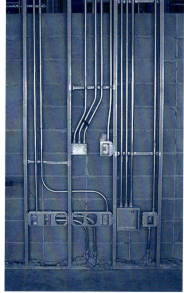

Site prep and rough-in can proceed once shop drawings are completed and approved. Rough-in includes, but is not limited to:

- Installation and/or confirmation of conduit, cable trays, risers and other cable management sizes and locations
- Installation of all cables as possible ("cable pulls")
- Structural mounting, blocking and other supports for ceiling-mounted devices (projectors, cameras, loudspeakers, monitors, etc.). Note that all structural mounts and systems should be designed for five times the anticipated loads. After installation, mountings should be nominally tested for load bearing and flex.
- Installation of projection screens (screens may be installed by either the AV contractor or the GC)
- Installation and/or coordination of mounting systems for wall-mounted displays (plasma, LCD); includes wall material, finish/treatment, blocking, location of electrical receptacles and junction boxes
- Miscellaneous junction boxes and other connection points and housings
- Monitoring of job site conditions and associated work by others. This includes AV conduit runs, floorbox and backbox locations, AV device mounting locations, data and power outlet locations, lighting control interface installation and other AV-related infrastructure items.

During any of these activities, all work should be coordinated with the other trades on-site. Intrusive and noisy work should be coordinated with the owner and/or existing occupants of adjacent areas.

Pre-Test Cable Systems

When pulling cables, all cables should be marked with the cable tag reference as shown on the shop drawings. These tags, placed at both ends of each cable, may be replaced with permanent cable markers when cables are cut and terminated permanently. Self-adhesive, printed cable markers are best for this use.

Once cables are pulled, they should be tested for continuity and integrity. This assures that all cables are properly labeled and have not been damaged or internally shorted. It is far easier to re-pull a cable during the rough-in process than when the equipment is installed and room finishes have been applied.

Sub-contractors to the integrator, GC, EC or owner sometimes perform the AV cable installation outside the AV contract. In this case, a report should be prepared by the cable installer and submitted to the integrator. This report should contain a list of all cables and the test results for each. Depending on the application and signal type, the integrator may require test results that include identification of each cable, continuity and conductor mapping, resistance, impedance, length and shielding.

Occupational Safety On-Site – Keeping Employees Safe

Ensuring that a job site is free of conditions hazardous to employees is everyone's responsibility. Employees must receive proper protection from any construction activities that require special chemicals (e.g., concrete etching) or pose physical hazards (such as welding, grinding, overhead work, and any activities that produce airborne pollutants or other physical danger). A few activities routinely performed by AV technicians may also create potentially hazardous conditions, such as soldering or structural work.

Climate is also an issue. Although some trades can function in temperature extremes, most AV work must be performed in a normal room temperature range. OSHA recommends temperature control in the range of 68-76° F and humidity control in the range of 20%-60%.[24]

The following are some typical job site safety precautions:
- Use of hard hats, safety glasses, protective clothing and footwear as required by site policy and/or OSHA regulations
- Provision of proper ventilation and HVAC control
- Provision of MSDS (Material Safety Data Sheets) on-site for all hazardous chemical and materials (to be provided by supplier and user of material per federal law)[25]
- Use of proper ladders, staging, lifts and supports
- Adherence to any and all federal, state and local laws/regulations
- Adherence to building and/or company safety policies

Step Five: Pre-Assembling and Testing the AV System

In most cases, systems are pre-assembled, wired and tested at a location other than the site of final installation — typically at the integrator's shop. This process allows assembly and testing in a controlled environment with appropriate installation materials and tools readily at hand. Part or all of certain systems must be built on-site, such as items to be installed in custom millwork (such as podiums) and retro-fit of existing customer's equipment with minimal downtime. Whenever possible, systems and sub-systems should be built and tested off-site.

Systems and/or racks should be assembled per the shop drawings, carefully following all notations and details. If the fabricator discovers that the actual assembly must vary from the drawings, approval by the system designer and/or consultant should be obtained prior to any change that affects system equipment, function or performance. No matter how minor, any deviation from the shop drawings should be noted and returned to the drafter for inclusion in the as-built drawings.

Often, AV equipment must be installed in other construction or furnishings prior to shipment to the site. For example, a custom millwork shop may require samples of table-mounted devices, such as microphones, boxes and connection panels. The integrator must make these goods or samples available on the other party's schedule.

Software Development and Testing

A number of software elements in modern AV systems must be programmed by the AV providers and often must be coordinated with the IT systems. Software development is required for a variety of AV system components including:
- AV control systems
- Audio and video DSP systems
- Surround sound processors
- Video and imaging processors

Corporate privacy and government security on-site

Owner and end-user privacy must be a top priority. During many construction and installation activities, project team members may witness activities that are germane to the owner's business. In whatever manner it occurs, intellectual property must be kept confidential by all project personnel. It is appropriate for the owner to require all project personnel to sign non-disclosure agreements that ensure the protection of intellectual property.

A company may also require proper identification and background checks of all personnel and request badges or other identification (including photos). In all cases, a requirement for ID must be non-discriminatory and conform to all laws. There may be other requirements determined by special government/intelligence security rules or agency policies.

---

[24] OSHA Standard #1910.1000 (interpretation): Go to www.osha.gov and perform an advanced search for documents with "humidity" in the title while limiting the search to "Standard Interpretations" only.

[25] OSHA Standard #1910.1200

Although some software may be created during the design process (such as DSP programming and touchscreen GUI designs), any required software should be tested to confirm that it meets the system functionality requirements according to the system design.

A third-party programmer may be sub-contracted by the owner, consultant or integrator to develop the AV-related software. The programmer must work with the integration team to load and test the software in the appropriate devices and resolve as many functional or graphical issues as possible during off-site testing. The programmer must also work within the integrator's delivery schedule for final installation, testing and training.

Shop Testing

All equipment and systems need to be tested in the shop prior to delivery and installation at the job site. Shop testing not only assures that systems function as intended, but potentially saves significant time in the field.

Equipment and all elements of the system should be tested to be sure that they meet manufacturers' and project specifications. Items not meeting criteria should be replaced immediately.

During shop testing, all components and signal paths must be tested and the results documented. The recommended range of tests includes:

- **Proper function of all equipment** — Each manufactured (stock or custom) item should be checked to assure that it meets published and/or intended specification and function unto itself. Transports should be checked for operation, recorders for quality recording, and all inputs and outputs verified for function. All inputs and outputs of switchers and matrix switchers should be checked for function and signal quality.
- **Mock-up of peripheral equipment** — Whenever possible, all peripheral equipment to be connected to the fabricated system(s), should be connected and checked as well. If the item to be installed is not practically available, a substitute should be found. Peripheral equipment includes loudspeakers, displays, cameras, microphones and other equipment that may not be mounted in an AV equipment rack.
- **Loading and testing of all software** — All software, GUIs, DSP code, drivers and others should be loaded and tested in the shop. Programmer(s) should be scheduled to be present to make any adjustments or corrections as required.
- **Pre-set adjustments** — During shop testing, all gain structures, levels and other adjustments should be optimized whenever possible. This includes input leveling, pre amp settings, peaking and equalization. Obviously, some settings cannot be finally set until installed at the site, but setting adjustments within the pre-fabricated system saves effort and time in the field.

It is appropriate (and sometimes required by contract) that an owner's representative and/or consultant to be present during shop testing. This provides not only an opportunity for verifying that the system is ready to go to the site, but also serves as a venue for familiarization with the systems prior to site delivery.

Packing for Shipment

Goods and pre-fabricated systems may be transported to the installation site by the integrator's truck, common carrier or a contracted transport service. AV systems are fragile and should be packed appropriately for shipment. Whenever possible, original packing should be used for items shipped separately; crating services are available from rack manufacturers and local carriers for long-distance delivery.

Step Six: Site Installation

Once the previous steps have been completed, the AV equipment is ready to be shipped to the site. The system to be shipped typically consists of rack-mounted equipment that has been pre-tested, peripheral equipment that must be mounted to walls, ceilings or furnishings, and portable equipment to be stored by the owner and used as needed once the system becomes operational.

Get GUI approval before programming

Shipping Equipment to the Site

The AV equipment should not be shipped to the site until the following conditions have been met:
- The site is prepped and "broom clean" i.e., all carpentry, drywall, paint, flooring and finishes are complete and the site is dirt- and dust-free. Dirt and dust are leading causes of premature equipment failure.
- All pre-fabricated systems have been completed and tested.
- All site-specific schedules have been checked — loading docks, elevators, occupied areas and security may all have specific schedules that must be honored.
- Security of the equipment can be guaranteed.

Once on-site, all equipment should be protected from any remaining hazards. Hazards include:
- Water damage, whether from construction activity, accidental water supply rupture or sprinkler system malfunction
- Physical damage from other construction activities
- Exposure to temperature extremes
- Exposure to excessive vibration
- Interruption of electric service and HVAC operation during equipment use

Safekeeping of the Equipment On-Site

By nature, a construction site is an open, unsecured area. Even with tight perimeter security, the sheer quantity and mass of the goods that are brought in and removed daily create a situation ripe for theft of AV equipment.

Although proper insurance, bonding, or official transfer of equipment to the owner is prudent and often necessary, the best preventive measure against theft is physical security. With a pre-assigned, central, partitioned AV area or control room, security is less complicated. For equipment located in open or other unsecured areas, security is more of a challenge.

There should always be an area designated for secure storage. When possible, this area should have a unique key with access given only to the AV integrator and other senior project or security personnel.

Under the design-bid-build process, a touchscreen GUI should be designed and approved during the design phase. For the design-build process, the GUI design may come a bit later in the overall project schedule. In either case, it is important to get approval for how the GUI looks and functions before developing the "back-end" programming that actually sends commands to the equipment.

Where possible, the end-users should be included in the review and approval process, as well as the AV consultant under design-bid-build. The GUI submittal may be in many formats, including HTML, PowerPoint, web or proprietary control manufacturer formats.

Plan for AV equipment security

Making a provision for creation of a secure area should be in the contract. Often the safest way is to wait until the owner takes delivery of the entire site before creating the "safety zone." Though environmental conditions might prevent it, external storage strictly under the integrator's control is another option (e.g., trailer). In any case, an equipment security plan should be made before delivery of any equipment to the site.

Site Installation and Pre-Commissioning Tests

Installation work should follow the technical guidelines defined in the project specification and shop drawings. An appropriate amount of supervision and technical resources should be on-site to allow a smooth and continuous installation process.

At the site, while many installation activities are performed solely by the integrator, some activities do need to be coordinated with other trades. These include mounting devices to the ceiling, wall and/or table and connecting racks and other equipment to building power.

As the installation nears completion, preliminary testing must be performed before system commissioning to determine if the system is substantially complete and proper connections have been made. While this preliminary test is often overlooked, it should not be. Testing can save valuable time during commissioning which in turn saves the cost of labor and re-visits for both the integrator and the AV consultant.

The preliminary tests before commissioning begins should include:
- Power on all equipment and verify the intended functions
- Verify signal paths for all field terminated wiring
- Adjust and align all displays for color, contrast and geometry
- Verify communications services (telephone, ISDN, Ethernet, and others)
- Configure and test the functionality of all audio- and video-conferencing systems
- When all testing is completed, a preliminary report (based on any format that may be prescribed in the contract) should be prepared indicating the results

Step Seven: Finalizing the Documentation

The final project documentation package should include the following.
- As-built drawings based on the shop drawings, reflecting all changes made to the system during both shop fabrication and site installation
- Architectural, mechanical, electrical and other utility changes that may have occurred during construction
- Dimensional data from field-determined or verified device locations (e.g., displays)
- One-line system drawing
- Schedule and documentation of all physical system settings and adjustments
- Gain settings, input attenuators and output settings
- Documentation of software configurations and settings
- Documentation of all DSP and other software configurations and settings
- Manufacturers' users guides, alphabetized, bound in 3-ring binders with index
- System-specific custom operation guide

Multiple copies of final documentation are normally needed and/or required by contract. Minimally, one copy should be available at the system location to aid in operation and trouble shooting. Additional copies should be provided as needed. Distribution in electronic format such as Portable Document Format (PDF) is usually preferred.

In addition, extra copies of the system-specific operation guide should also be available as handouts during any training session(s). A complete, draft version of final documentation should be available for reference (and/or editing) during system commissioning. Once system commissioning is complete and documentation is final, distribution copies may be produced.

Deliverables for documentation should include at least one printed copy for use and quick reference at the project site. Electronic copies should be made available for all end-users and per contractual requirements.

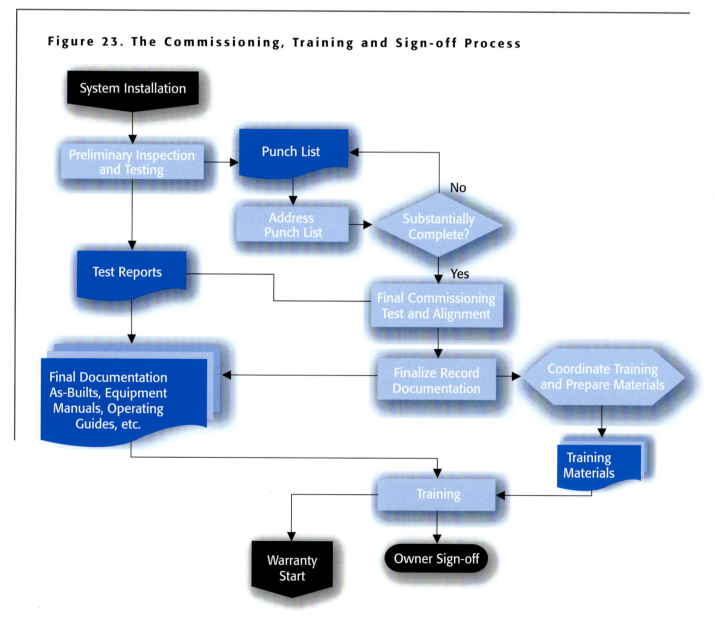

Figure 23. The Commissioning, Training and Sign-off Process

Chapter 6 — The Construction Phase

Chapter 6 Checklists

AV CONSTRUCTION KICKOFF MEETING AGENDA

Once the AV integrator is contracted for system installation, a construction/installation kick-off meeting (sometimes called a pre-construction meeting) should be held to get everyone oriented to the project and the AV issues that will affect the construction process. This includes the general items of site access and administrative procedures as well as the AV-specific issues. This sample agenda contains a comprehensive list that would typically apply to larger projects, but all of these items should be considered for any size project.

- ❏ Project and Meeting Participant Introductions
- ❏ Project Goals and General Scope of Work for AV

Project Communications
- ❏ Relationship of AV integrator and General Contractor
- ❏ Shop Drawings and Submittal Procedures

Site Access
- ❏ Allowed work and access hours
- ❏ Access and sensitivity to nearby existing occupied owner spaces
- ❏ Parking
- ❏ Boundaries of the project
- ❏ Security clearances and ID badges
- ❏ Keys
- ❏ Temporary office space.

Deliveries and Storage
- ❏ Loading dock location
- ❏ Hoist/Elevator size and schedule
- ❏ Equipment and rack load-in pathways
- ❏ Onsite Storage
- ❏ Offsite storage - warehousing and bonding
- ❏ Union and trade organization agreements

Working at the Site
- ❏ Limitations on noise- and vibration-producing activities
- ❏ Other events at the site
  Services/Utilities for site work
  - ❏ Toilets
  - ❏ Telephone/Office
  - ❏ Water
  - ❏ Heat
  - ❏ Electricity
  - ❏ Shutdown of Services Procedure
- ❏ Signage
- ❏ Housekeeping

Hazardous Materials at the Site
- ❏ Asbestos
- ❏ Lead
- ❏ PCB's
- ❏ Material Safety Data Sheet (MSDS)

Changes in the Work
- ❏ Requests for Information
- ❏ Change Order Requests
- ❏ Change Orders
- ❏ Known addenda or change orders to date

AV Coordination Issues During Construction

General Contractor
- ❏ Rough openings by GC for projection screens, loudspeakers, racks and other devices
- ❏ Items provided by AV integrator and installed by the GC
- ❏ Structural support being provided for AV devices such as projectors and loudspeakers
- ❏ AV-related furniture, casework and millwork construction, delivery and coordination
- ❏ Room and plate finish coordination
- ❏ Ceiling Installation schedule

Electrical
- ❏ Importance of backbox and floorbox locations
- ❏ Conduit, backbox and floor box installation schedule
- ❏ Special issues for AV-designated power outlets and junction boxes (TVSS, Isolated ground)
- ❏ Loudspeaker backboxes, floor boxes or other devices provided by AV integrator but installed by EC
- ❏ Power activation for AV systems
- ❏ Lighting systems fixtures, dimming and interface installation

Other Low Voltage Trades
- ❏ Data/telecom contractor scope and schedule (structured cabling installation, cable TV infrastructure and other related scopes of work)
- ❏ Cable coordination with other trades (pulling schedules, sharing of cable trays, patch bays, floor boxes, etc.)

HVAC
- ❏ HVAC duct and device locations and install schedule

Contract Billing and Budget
- ❏ Processing of contract documents and pay requests
- ❏ Any potential budget/scope issues or changes

AV Contract Project Schedule
- ❏ Notice to Proceed Date
- ❏ Start Date
- ❏ Shop drawing submittals and approvals
- ❏ Cable pulling and other on site work
- ❏ Delivery of equipment to be installed by others
- ❏ Required receipt of Owner-Furnished Equipment (OFE) from owner
- ❏ AV equipment delivery and installation
- ❏ Preliminary test and checkout
- ❏ Substantial completion date
- ❏ Final commissioning and checkout
- ❏ User training
- ❏ Completion date
- ❏ First AV event date

Owner Issues
- ❏ OFE Delivery
- ❏ Owner data/telecom Issues
    - ❏ Ethernet connectivity, used for communications by control systems, video conference, projectors, and many other AV devices and systems
    - ❏ Subscription and activation of external and internal data/telecom services (PRI/BRI, T1/2/3, ISDN, analog)
    - ❏ Dial-up services, including ISDN, BRI/PRI, analog telco (a.k.a. POTS), primarily for conferencing use
    - ❏ IP addressing and subnet schemes
    - ❏ Network segmenting, routing, and VLAN/VPN schemes
    - ❏ Network services startup schedule

- ❏ Activation of electrical service and other utilities
- ❏ Activation of cable TV and satellite subscription services
- ❏ Communicate and monitor adherence to building/company standards and policies
- ❏ Provide access to existing building information in renovation/restoration projects
- ❏ End-User move-in and first use of the AV system

Future progress meetings
- ❏ Schedule and attendance

Facility Tour
- ❏ Tour existing facility

AV CONSTRUCTION KICKOFF MEETING AGENDA

– Continued

SITE READINESS AND SECURITY CHECKLIST

Before delivery of AV equipment and racks for field installation, the items in this checklist must be completed to provide an acceptable environment for installation of the AV systems on site.

Hazardous conditions
- ❏ All water piping completely installed and tested (supply, HVAC, drains, fire protection, etc.)
- ❏ Dirt and dirt sources removed. AV-rooms at least "broom-clean".
- ❏ Dust and dust sources removed. HVAC running with filters in place.

Major construction activities completed, especially activities that may create physical damage to equipment or racks such as:
- ❏ Overhead work that may cause debris or dust
- ❏ Chemical work such as concrete cleaning and finishing
- ❏ Welding
- ❏ Grinding
- ❏ Activities that may cause excessive vibration

- ❏ Site occupational safety procedures in place and enforced

Security
AV areas established and secure for
- ❏ Equipment, rack and tool storage
- ❏ Staging and work area for final assembly work on racks or in furniture and millwork
- ❏ Equipment installation areas that are not yet turned over to the owner
- ❏ Keys provided to AV integrator with limited distribution
- ❏ Awareness and/or required training obtained for building and/or company security policies
- ❏ Security badges and clearances obtained for all on-site personnel
- ❏ Corporate Privacy or Government Security non-disclosure agreements properly executed

Infrastructure
- ❏ All required AV conduit in place
- ❏ AV field cabling pulls complete
- ❏ All AV-related AC power wiring in place.
- ❏ AC power on, tested and reliable
- ❏ Screen rough-in and other physical preparations for AV verified
- ❏ Floor box and backbox locations verified
- ❏ Owner Furnished Equipment (OFE) provided
- ❏ All owner supplied millwork and furniture available
- ❏ System interfaces by others installed (lighting, screens, motorized drapes, other)
- ❏ Cabling by others in place (data/telecom, cable TV)
- ❏ Lighting installed and available for use in AV spaces

Physical Environment
- ❏ Humidity and temperature properly controlled by HVAC
- ❏ Floor finish installation complete
- ❏ Ceiling work complete except for minor punch list items. Ceiling tile in place in lay-in ceilings
- ❏ Wall construction complete and finishes installed (paint, wallpaper, fabric, acoustical panels, etc.)
- ❏ Structural backboards and other AV mounting infrastructure installed and verified
- ❏ Load-in path for racks and equipment established and verified

Title and Legend
- ❏ Title page including sheet index
- ❏ Legend of symbols and conventions

Architectural Plans and Details
- ❏ AV device location and coordination floor and ceiling plans showing room-mounted device locations, rack layouts, seating layouts and critical dimensions
- ❏ Large scale plans of important, high-density areas such as equipment and control rooms.
- ❏ Millwork and furniture integration details for AV equipment

AV Electrical Floor, Wall and Ceiling Plans as Needed
- ❏ Conduit risers and plans
- ❏ Junction and backbox location plans including cable fills
- ❏ AV electrical requirements showing AV-related power outlets and junction boxes
- ❏ Voice/data, CATV, other requirements showing outlet locations and complete system if provided under the AV contract

Cabling Details
- ❏ Cable Run List (may be a separate document) including cable types, quantities, estimated length, wire ID, use
- ❏ Cable/connector termination details

AV Elevations and Mounting Details
- ❏ Direct view display device (flat panel and CRT) mounting and viewing angle details
- ❏ Video projector mounting details including plan, section, projection paths and viewing angles
- ❏ Loudspeaker mounting and aiming details

Equipment racks
- ❏ Front, rear and side elevations including device locations and wire management scheme
- ❏ Security/locks
- ❏ Ventilation
- ❏ Details

AV Signal Flow Diagrams
Full signal flow diagrams for audio, video, and control and AV power including:
- ❏ Complete equipment interconnection information
- ❏ Wire Cable types, manufacturers, model
- ❏ Cable number/label assignment for every cable
- ❏ Device manufacturer and model numbers
- ❏ Any device labeling used (such as DVD-1, VCR-2, etc.)
- ❏ Device and rack locations
- ❏ Full wire/connect fan-out and device connector details
- ❏ Any critical equipment settings that could cause damage or affect performance of the system at startup such as amplifier mode, control system IP addresses and video signal format settings

- ❏ One line flow diagrams for less cluttered and more easily navigable views of large or complicated systems.

SHOP AND AS-BUILT DRAWING COMPONENTS CHECKLIST

A shop drawing should be a complete graphical description of the AV system. It is augmented and complemented by the equipment and operations manuals and other materials to create the final documentation set. The difference between the shop drawings and as-builts is that the as-builts are corrected to reflect the actual installation configuration which may have changed from the original shop drawings. The as-builts, in turn, become part of the record drawings for the project contractually.

Additional documentation is normally required for complete submittals and as-builts. See the Final AV System Record Documentation Package Checklist for the full list of typical record documentation package components.

SHOP AND AS-BUILT DRAWING COMPONENTS CHECKLIST

— Continued

AV Detail Drawings
- ❏ Connection plate details
- ❏ Device and signal grounding between devices and at patch bays
- ❏ Terminal blocks
- ❏ Custom circuitry
- ❏ Pinout schemes for all connectors
- ❏ Communications interface wiring
- ❏ Ventilation details
- ❏ Miscellaneous details of podiums, rack cavities, etc.

CHAPTER 7
System Commissioning and Training

Step One:	Perform Preliminary Tests on the Completed System Installation
Step Two:	Generate the Punch List
Step Three:	Establish Substantial Completion
Step Four:	Inspect, Test and Align
Step Five:	Train the Users
Step Six:	Sign-Off and Start the Warranty Period

CHAPTER SEVEN

As oft-quoted New York Yankees catcher Yogi Berra once said, "It ain't over 'til it's over," and certainly the AV project is not over until the system commissioning phase is complete and training has been provided to the users.

Throughout the processes described in this book, many aspects of the AV integration process are focused on the AV infrastructure design, installation and coordination. With this work done and the systems installed, the commissioning process focuses on the system itself, on its electronics and on its functionality within the environment that has been created.

System commissioning plays a pivotal role in the entire audiovisual project. The commissioning agent ensures that standards have been followed, verifies that contractual obligations have been met, and checks that the system is ready to perform properly in its intended use. All aspects of the system are tested, adjusted and optimized. Workmanship is inspected, system performance is evaluated, and deficiencies are identified and corrected.

After the commissioning process is complete, training is the next step to complete the handoff to the end-users. The end-users, both technical and non-technical, must receive instruction to ensure the ongoing success of the facility and AV system. Minimally, the end-users and those responsible for operating the facility need adequate training on day-to-day usage and ways to address a system that does not function as expected.

Beyond training, there are often other post-commissioning tasks associated with the AV systems contract, such as preventive maintenance, warranty support and additional training for new users, as well as refresher training once the system has been used for a period of time.

The Basics of System Commissioning

In general, the commissioning process assures that the design intent has been met. To this end, both objective and subjective tests are performed, using the design specification, functional intent and industry standards as metrics against which the completed system is compared. Each sub-system and coordination item is inspected, tested and documented. In addition to touching on the technical components of the commissioning process, this chapter focuses on the process itself. The two basic concepts of system commissioning are:

> **Verification** of the quantitative aspects of the system and the individual components. This includes verification that all electronic components have been provided and installed, and are functioning at or beyond their specifications and according to the design intent.
>
> **Optimization** of the overall system's functionality and quality. This includes aspects of the system such as video projection alignment, audio quality and control system operation.

These two fundamental principles drive the process and determine what tasks are required to complete it. The overall process has four steps:

1. Preparing for commissioning
2. Generating a punch list
3. Establishing substantial completion
4. Inspecting and testing the system

All the steps are geared to the overall goals of verification and optimization. Verifying that the design intent has been met and optimizing the operation of the AV systems translates into specific tasks for the commissioning process such as:

- Verify and optimize audio and video signal paths throughout the system
- Verify that control systems and user interfaces are operating correctly and efficiently and provide the required functionality
- Complete all programming of audio and video devices and verify their functionality
- Verify that all contractual obligations have been met, including the complete system installation and provision of documentation

Successful completion of these tasks will close out the installation process and provide the basis for conclusion of the installation contract.

Although testing has been performed on the systems at various stages of the process, the commissioning process is specifically designed to uncover problems. Many problems that arise during this phase are easily solved and can be addressed by the AV team. These typically range from wiring mistakes and incomplete labeling to equipment failures and programming errors.. These issues are considered punch list items under the contract.

Other issues that emerge may be the result of user requests, a change in end-user requirements, or technology that has changed since the project was designed and contracted. These issues may be outside the original scope of work. In some cases, these changes can be accommodated without a change to the contract, while in others a change order or separate contract may be required.

A commissioning report should be produced that details problems and identifies any punch list items still outstanding, and if they will be addressed at a later date, a new agreement should be drawn up covering the future work.

Who Commissions the System?

AV system commissioning requires significant expertise in audio, video and control systems design and operation, as well as familiarity with the project being commissioned. Individuals designated to perform commissioning will vary from project to project.

For the most objective evaluation, an experienced professional other than the installation technician(s) should perform the system commissioning, though the technicians often assist with the commissioning tasks.[26] In theory, any person or team technically competent in pro-AV can serve as commissioning agent. Typical scenarios include:

Consultant-Led Project: The AV consultant typically leads the system commissioning process. The integrator and consultant should work as a team to perform evaluation and adjustments needed to finalize the system. The contract often specifies a minimum number of hours for the integrator's personnel to assist with commissioning. The owner may also participate in the commissioning process.

Integrator-Led Project: While the field installation team may assist with the process, the leader of the commissioning process is typically a senior-level technical resource who is familiar with the required test equipment, industry standards, and performance criteria. The professional should be familiar with the project if it is especially large or complex. It is recommended that an owner designate an employee or other independent resource to assist and/or witness the system performance evaluation.

Owner-Led Project: If the owner has provided the AV design or has the internal expertise to provide a proper evaluation of the systems, a member of the owner's AV team may lead the process. In some cases, the owner may elect to contract with a third-party AV provider who offers commissioning services.

Third-Party Agent: On some construction projects, a commissioning agent is contracted to commission some or all of the building's systems, which may include the AV systems. This is typically done with the aid of a subcontracted AV provider other than the one who installed the project.

In all cases, the senior installation technician should be present to assist with commissioning since he is familiar with the system. The tech should be prepared with appropriate test equipment, tools required for adjustments and minor corrections and a laptop computer if required for adjustment to AV software configurations such as control systems programming, DSP devices or AV-related data network configurations. If the system includes a control system, the programmer should also be available and ready to make programming corrections.

When to Commission the System

Ideally, system commissioning commences when the AV system installation is complete and all documentation is ready. Attempting commissioning before the system is complete can result in wasted time, a long punch list of incomplete work, and additional costs for the commissioning entity to return to the site upon completion of installation.

Conversely, commissioning long after the system is complete is not recommended since it may delay the warranty start date, contract closeout and final payment to the integrator. The rule of thumb in this phase is "be ready, and don't delay."

---

[26] In very small installations, the installer may have to commission the system. If so, the installer must be objective and follow the original design specification and functional description as a reference to establishing compliance. A conscious effort must be made to evaluate the system with the best interest of the client or end-user in mind.

Step One: Perform Preliminary Tests on the Completed System Installation

Prior to system commissioning, the AV systems should be checked for basic alignment and functionality. This will prepare the system for the final commissioning, allowing the final checkout to focus on the qualitative aspects of the system.

This preliminary check may be included in the contract. Typically, it involves the installation technicians or other technical resource designated by the integrator. Under a consultant- or owner-led option the AV designer may be involved in a preliminary checkout before the final commissioning. In this case, the integrator should perform any appropriate testing and verification before the consultant or owner arrives for this site visit.

The preliminary checkout and alignment includes a wide variety of tasks such as:
- Verification that all equipment (purchased or OFE) has been delivered and installed per specifications, and that a list of all equipment, complete with model and serial numbers, is available
- Verification that other trades have completed the work associated with the proper functioning of AV systems
- Confirmation that communications and networking services are installed and working (Ethernet, IP addressing, ISDN, telephone, and other applicable services)
- Testing of signal paths for all video and audio systems
- Programming and testing of conferencing systems, including verification of IP addressing, SPID identification, port/jack numbers for required services and other connection identifiers
- Verification of successful control system operation
- Adjustment of the audio system for proper operation, including any DSP programming
- Testing of wireless microphones, including confirmation of transmitter/receiver frequencies, channels, range and charging of batteries
- Testing and alignment of video monitors and projectors for good image reproduction from all potential sources

This is the stage where field wiring problems, improper physical alignment of speakers and projectors, and previously undetected equipment failures should be discovered and corrected.

Allowing Time for Evaluation

Adequate time should be planned for performance evaluation. Smaller systems can often be tested in half a day. Larger, more complex systems may take several days. Especially large or complex projects may require a phased schedule over several weeks to commission.

Picking the right time to conduct the performance evaluation should be a joint decision of the integrator, owner and consultant (if applicable). Contractual obligations, the work of the other trades, and use of the facility must be considered. Testing, evaluation and alignment tasks should be considered in the project schedule from the beginning of the project, since these tasks usually occur after the owner has begun to occupy the facility. There may be a need to resolve scheduling conflicts or to address end-users' desire to use systems that are not yet ready for operation.

Assessing multiple, identical systems and mock-ups

When many identical systems are to be installed (e.g., a classroom building with many identical classrooms), it is preferable to build one system to completion and evaluate it early in the project. This serves as a model, or prototype system, in which design issues and corrections can be made before they are duplicated. Once the prototype has been optimized, integration refinements – both minor and major – can be applied to all subsequent builds.

Occasionally, the overall scope of the project may warrant assembly of a mock-up before the site is ready for installation. This mock-up allows designers and end-users to evaluate the planned system, sometimes requiring changes to the original design. Mock-ups often include not only AV equipment but also full-size models of teaching stations, conference table sections and lecterns. The mock-up may also be used for training purposes. When the actual systems are built, the equipment, if still in as-new condition, may be used. Alternatively, equipment used in a mock-up may be stored as a spare or simply declared as surplus for use elsewhere.

As in the earlier phases, time often means money. There will be costs associated with the commissioning process because it can involve staff from the owner, integrator, consultant and third-party commissioning agent depending on the project configuration. Commissioning can represent a significant cost-per-hour. If the system is not ready when the commissioning agent arrives, the integrator may be held responsible for the time and expenses of others related to additional visits, costs to the owner or the integrator's internal costs required to complete the process. If there are extenuating circumstances contributing to a delay in testing, these should be communicated, agreed to and negotiated prior to scheduling commissioning.

Step Two: Generate the Punch List

The punch list is a key element in the project process, because it becomes the final checklist for a complete installation and contract closeout. Depending on the contract language and relationships, the punch list may be created by the AV consultant, the owner's AV project manager, or the AV integrator's project manager or other internal personnel. The preliminary punch list may be internal to the integrator under many design-build projects, but is usually required for distribution to various design and owner team members under most other methods. The "final" punch list that is generated after the final commissioning test and alignment is usually distributed to the designated project team stakeholders under any method.

During the preliminary checkout outlined in the previous step, a preliminary punch list should be developed that includes all system deficiencies discovered, along with the possible resolution of each deficiency, and the party responsible for each item. This punch list should be distributed to the responsible parties for completion, and should include due dates for completion of each item.

Each punch list is unique to the project for which it is generated, but some typical causes for items that may appear on a punch list are:
- Poor AV connector terminations
- Damaged wiring
- Workmanship issues with equipment installation or aesthetic components of the work such as damaged wall finishes, undesirable cable management and other problems that are visibly objectionable
- Physical installation issues such as projector positioning, loudspeaker locations and alignment and integration of devices into furniture
- Delays in delivery of AV equipment
- AV equipment failures
- Slow delivery or non-delivery of goods by non-AV service providers, e.g., millwork, electrical and other contractors
- Slow or non-delivery of OFE or communications and network services

While some items are the direct responsibility of the integrator, some are caused by delay in work by other parties. To resolve these issues, the AV integrator must play an active role in seeking timely solutions from the other parties. Proper planning, documentation, and communication are crucial.

Step Three: Establish Substantial Completion

In the normal course of the project sequence, the integrator completes the installation and conducts preliminary testing to verify that the system components are properly connected and operational. On large projects, a preliminary test may require the participation of others besides the integrator, and a report may be required for distribution documenting that:

- The preliminary tests have been completed. This should include the results of the preliminary tests, any exceptions to be addressed and the anticipated resolution. If not previously reported, any shop test results should also be included.
- The system installation is complete per the contract (including change orders) pending final commissioning. Any punch list items that have been developed to date should also be included.
- The system is ready for final testing and alignment.

In most cases, the accurate completion of these tasks and the distribution of the report documenting these items establishes what is generally known as "substantial completion." This term is often used in construction contracts to indicate that most or all of the work is completed with the exception of some punch list items. The substantial completion date is sometimes used as the start date of the warranty, while in other cases, the warranty may not be started until after commissioning is complete. Warranty start date should be explicitly called out in the contract documents. If not, the warranty start date may be negotiated during the commissioning phase, although this is less desirable than establishing it up-front..

For smaller systems, establishing substantial completion may be relatively simple, while in larger projects it can become complex. Sometimes different areas of a large project may "come on-line" at different times, and substantial completion as well as commissioning may need to be on a staggered schedule. This can sometimes require multiple site visits, multiple training sessions, and variations in the warranty start date for systems in different areas of the same project.

The integrator submits the substantial completion report to the commissioning agent, following the submittal method established previously under the contract. After review, all parties schedule a time to perform system commissioning.

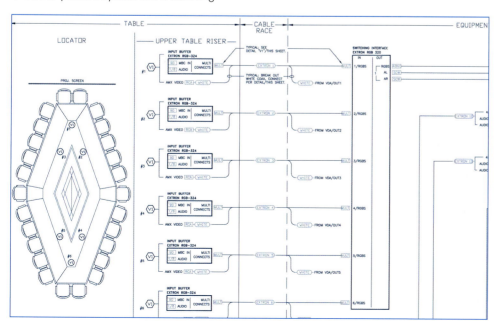

Step Four: Inspect, Test and Align

The inspection, testing and alignment of the AV system are the essential tasks of commissioning. The premise behind commissioning may vary from project to project. If a third party agent who has not been a part of the design and installation team conducts the commissioning, the preliminary checkout and alignment and the resulting report may need to be more thorough. If the commissioning is performed by part of the AV project team, certain steps can be condensed; for example, the preliminary report may not need to include a record of all equipment settings, as that will be provided by the team as a part of the final commissioning report.

A copy of any criteria established during system design and specification, as well as the preliminary test report, system and equipment documentation, as-built drawings and software, should be made available for reference and use at the testing and alignment site visit.

Facility Inspection

Before beginning tests, it is useful to tour the facility and become familiar with physical locations, meet new faces and conduct a last-minute check of the facility schedule. The condition of the site and completed installation work can also be reviewed. A final determination can be made as to whether or not the facility is ready for the commissioning process.

Nominal inspection can also be made of physical locations, mounting location of portable equipment and cables, room numbers (which may have changed during a construction project), and power and lighting sources. This early inspection will save time during testing, as well as provide an opportunity to resolve physical issues before they delay the process.

Existing Documentation and Inventory

All documentation should be available in draft or near-final form during system commissioning. As-built drawings, one-line drawings, manufacturers' user guides, and custom-written manuals should be inspected for completeness and accuracy. If any of these materials are unavailable during commissioning, it should be stated in the Report of Substantial Completion and confirmed by all parties. In cases where crucial elements are not available, such as shop drawings (or as-builts) or software for DSP and control equipment, the facility should be deemed not ready for commissioning.

A physical equipment inventory should be taken and compared to all applicable documents: the specification requirements, contractual requirements, quote/bid response, submittals, change orders and integrator-prepared as-built equipment list (including serial numbers).

Testing the Systems

Objective testing is a key element of commissioning no matter what delivery method is used. Testing of system function and performance must be thorough, technically accurate and documented. Where applicable, industry standards for system performance should be used as a benchmark. Equipment needed to measure a system's performance includes signal generators, oscilloscopes, optical measurement devices, and audio spectrum analyzers. Test equipment must be operated by technically capable personnel and the findings recorded properly.

In almost all cases, the integration team is responsible for furnishing and operating this gear. Equipment should be properly calibrated and in good working order. Technical specs of test equipment should match the intended application or test.

The required test gear is often listed in a design specification created in a consultant-led project. If any piece of test gear is unavailable, the commissioning agent should be notified in advance. If it is necessary, equipment not available will have to be rented.

The nature of certain aspects of the AV system may require that they be evaluated subjectively. Though there are related component standards, evaluating a "good" solder joint, "good" cable management, and "good" interface design may require some subjectivity. The more detail and standards that can be incorporated into an AV contract, the more objective the evaluation can be concerning meeting the contract requirements.

In data/telecom structured cabling, industry-specific standards allow for a consistent pass/fail evaluation of almost every aspect of the work. At this point in time, certain aspects of AV installation have not yet established industry-level pass/fail standards due to the wide variety of issues that affect AV work.

However, most required AV tests and measurements can be made objectively using test equipment designed to evaluate the performance of audio, video and optical systems. And while thorough testing is required for every system, it is often impractical to evaluate the minutia of a very large system using objective instrumentation. Without question, the experience of the commissioning agent is crucial to a successful performance evaluation and system commissioning.

The items below represent an overview of the types of evaluation and testing that should be performed. Any or all of these tests may require adjustments in analog, digital, and control devices and programming.

Workmanship
During a performance evaluation, general workmanship should be examined for potential problems. For example, visual inspection of connectors can signal existing or future problems. Connectors should be made based on the manufacturer's design specification standards, with all covers, strain relief and fastenings installed.

Audio and Acoustic Systems
During checkout of the audio system, all connections, routing and levels should be checked and adjusted or corrected to verify and optimize the operation of the audio playback, speech reinforcement or other audio systems. These systems may include virtual DSP devices as well as the "real" devices. Acoustic testing can be performed to either verify that the acoustic environment is acceptable or troubleshoot an AV system problem that may be caused by an acoustical parameter.

Video and Optical Systems
As with audio, all video connections, routing, and levels should be checked and adjusted or corrected to verify and optimize the operation of video sources, processing and displays. All video switchers, routers, distribution amplifiers and signal processing equipment such as scalers, processors, scan converters, transcoders and multi-window processors must be checked for proper performance and adjusted as needed and applicable.

Chapter 7 — System Commissioning and Training

Network and Telecommunications Services

Though some network subsystems may be provided and installed by the AV integrator, most network and other telecom services aren't typically delivered as part of the AV system contract. In the cases of elements not provided by the AV integrator, there are connections that require testing to verify their operation and provisioning. Each of these systems, including LAN/Ethernet and dial-up services such as ISDN, PRI, T1, analog telephone and other communications services should be tested with the AV system operation. Problems discovered at this point may require additional testing and configuration by the integrator and/or the owner's staff or the communications service provider.

Conferencing Systems

All conferencing and web-based systems should be checked for proper function. In particular, making calls (audio and video) to a system of known quality with a qualified user on the far end is crucial to proper evaluation and optimization of these systems.

RF and Transmission Systems

RF systems such as cable TV and master antenna distribution systems (as delivered under the AV scope of work) should be checked, adjusted, and optimized for best reception of signals. There may be contractual crossover in this work since cable TV cabling (particularly horizontal cabling from distribution rooms) is often under the data/telecom contractor's scope of work. The headend system and riser cabling may be under the AV contract, so these items may need to be coordinated.

AV Infrastructure Systems

Certain systems primarily related to infrastructure that supports the AV systems are most often furnished or installed outside of the scope of the AV contract, but must work in harmony with the AV system. These systems typically include lighting systems, projection screens, motorized drapes and shades, operable walls and HVAC. For these systems, standalone operation and interoperability with the AV system (when specified) must be verified.

In many cases, adjustment of these systems must be coordinated with the owner or supplier of the system to be tested. For instance, the AV designer may be best suited to define the scene settings in a lighting system but must coordinate these settings with the EC or lighting designer responsible for the lighting system. In addition, settings for these supporting systems should be included in the AV documentation for reference.

Also, AV signals may be transported through methods furnished outside of the scope of the AV contract, such as fiber optics, microwave and satellite systems, requiring coordination of connection, testing and optimization with the supplier of these services.

User Interfaces and Control System Programming

One of the biggest challenges during the commissioning process can be the testing of control systems for user-friendliness and proper functionality. Functions and controls must be tested both individually and for their effect on one another. In a practical sense, each primary function should be tested from system startup, and then re-activated from all possible system states.

The control system software may be written by a member of the integration team, the consultant, the control manufacturer or a third-party programmer. In any event, the individual who programmed the system must be present during the performance evaluation to answer questions and, when practical, make ad-hoc programming corrections.

If the design specification includes a button-by-button functional description, these functions should be checked. Variations from the specification may be acceptable because of unforeseen equipment functions, programming issues or requests from the owner or AV designer. These should be resolved with the owner and the designer before being accepted.

The control system checkout can encompass many aspects of the AV system as well as other related systems in the building. Some of the elements to be checked are:

User Interfaces
User interfaces may include wired or wireless touchscreen and button panels, web interfaces and programming or stand-alone applications. These should all function as intuitively as possible and according to the design intent including all functions that were specified. Wireless devices must be checked for range and interference.

Connections to Other Systems
Connections to and controls for systems to be interfaced such as lighting, screens and shades need to be tested. Each should be checked for required functionality, interoperability and undesired interaction between systems. This is especially important when there are both system-specific wall-mounted controls and an AV system control interface for the same system.

Communications to Data Servers, e-Mail Systems and Pagers
Control systems may be programmed to report system status to a remote device or server for notification of technicians for service, help desk functions or logging of equipment use. Operation of these external systems should also be confirmed for configuration and proper function.

The Commissioning Report
Once the verification and optimization process is complete, the results should be documented and a recommendation made for the contracting entities (usually the owner and/or the general contractor) to accept the system as complete and commissioned. There may be lingering minor issues that still need to be addressed but are not significant enough to prevent acceptance of the system. In such instances, the course of action to address these issues usually needs joint agreement by the owner, end-users and the AV providers prior to recommendation for system acceptance by the owner.

The commissioning report can be delivered personally to the client or distributed through the project submittal process. Either way, it should contain:
- Project identification, scope, and specification
- Executive summary of findings
- Results of all objective tests. If voluminous, this may need to be either a paper-based appendix to the report or provided in electronic form, or both
- Results of all subjective tests
- Punch list of all incomplete work and/or items needing attention, including the course of action and dates for a resolution
- Description of any work, determined as needed, that is outside of original scope and contract
- Any other remaining issues to be resolved after commissioning
- A recommendation for system acceptance by the owner

The "last-minute" programming contingency

Some contracts may appropriately call for a "contingency" of hours to customize the control system during the commissioning process. Such contingencies do not apply to fixes due to problems with the software, but are intended to accommodate minor changes in the programming or interface if requested by the AV designer or the end-users. Any major changes beyond this contingency are needed, they would require contractual action to implement.

Step Five: Train the Users

The success of a completed audiovisual project lies not only in the systems design and execution, but also in the ability and willingness of the user to utilize the system to its full potential. Considering the often significant investment in the AV system itself, it is imperative that there be sufficient follow-through with the owner's personnel to ensure the system functions at its optimal level. This can only be accomplished through proper training and user documentation.

Comprehensive training of end-users and support personnel serves many purposes:
- Teaches the end-users how to use the system
- Improves the confidence of the end-users
- Reduces service calls resulting from "user error"
- Reduces chance of improper use – and subsequent damage of systems
- Enables in-house personnel to perform basic troubleshooting, possibly preventing lost use of system, and orients them when receiving assistance by telephone

The Training Program

In both a consultant- or integrator-led project, specific training needs may have been identified during the design phase and included in the contractual obligations. This may include duration, number of sessions, and anticipated number of trainees.

Figure 24. Sample Custom End-User Operation Guides

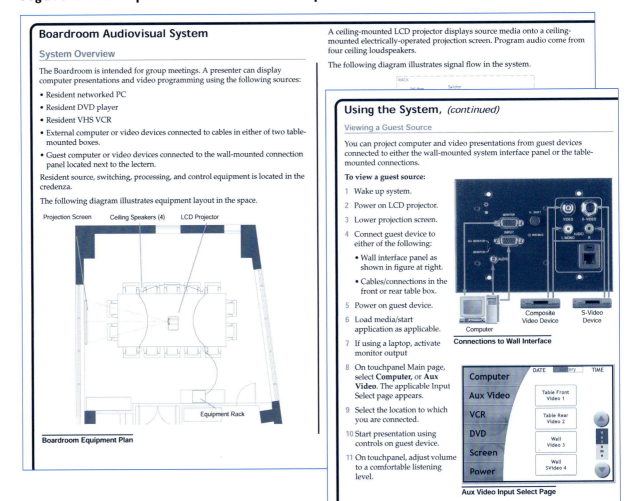

Operational documentation may also be included in the spec and/or contract, including system-specific custom-user guides, and manufacturers' owners and service manuals. Even if user documentation is not required by contract, training material still should be prepared. Simple and practical explanations of basic functions are best, often in the form of "flash cards." In the case of multiple systems, small laminated guides may also be appropriate.

All user and training documentation should be available during all training sessions.

When Should Training Start?

Whenever possible, training should be scheduled after the system is complete and all punch list items have been addressed. It is not practical to offer training on an incomplete system. However, on a large project with multiple systems in various rooms, training can be done on smaller stand-alone systems for users who do not need training on the overall project as individual systems are completed.

Sessions should be recorded to enable the owner to provide later training to those unable to attend, as well as for future training or review. Many recording methods exist, from a simple camcorder to media-rich PC/server-based systems. Sessions that will be recorded "live" with trainees should be planned and partially scripted accordingly. If time and cost permit, holding a separate session "for the camera" is ideal.

Who Needs Training?

A project may require several levels of training to cover various types of non-technical users as well as the AV technical staff. For example, someone who gives slide presentations in a media-rich space may need basic training on the operation of the system only. In the same space, another user may routinely host video-conferencing calls, using various media sources and cameras, requiring more advanced training. There may also be a media department or other technical staff who require detailed training for maintenance and troubleshooting of the system.

Prior to training, the trainers should know the background and abilities of the trainees in order to prepare suitable material. Depending on their knowledge, it is often prudent to include training on AV basics. Review of signal and connector types, best seating positions, nuances of various devices, and recognition and identification of improper function may also be appropriate. All of this should be done keeping the knowledge, background and interest of the audience for a particular session in mind.

It is usually most effective to train smaller groups (up to six people). This allows all participants some "hands-on" time. If there is an appropriate venue, larger groups can be accommodated; this may be the case for training instructors on systems in school classrooms.

In addition to end-user and technical training, it may be appropriate to "train the trainer," or teach select individuals how to train so that training can continue within the organization. This person should be familiar with the design philosophy, troubleshooting, and all possible scenarios in which the system might be used.

BEST PRACTICE

Develop a training department

For more sophisticated systems, some larger integrators have training departments that can put together a customized professional training program for each project with the help of the technical staff who worked on the project. Such a structure is an excellent benefit to the end-users and is highly recommended for integrators whose operations can support such a department.

In these cases, even though a professional trainer may lead the training session, the system designer and other technical integrator personnel who worked on the project should be in attendance. For training technical users, the system designer will usually be required to lead the session.

The same concept holds true for end-users. When an organization is large enough and uses a significant amount of AV, an internal training program (or an outsourced one) should be considered to keep end-users comfortable with the systems and accommodate new users entering the organization.

Who Does the Training?

The entity that handles the training can vary depending on the nature of the project delivery and the contractual arrangement. In general, it is best practice for the system designer to conduct or be a primary participant in the training. This may be the AV consultant or the AV integrator's designer depending on the delivery method chosen.

The system designer, who is privy to the program and is the leader of the design phase, is fully aware of the client's needs, as well as the intent and scope of the design. In fact, the designer often will have worked with the trainees during the program phase, and this continuity can be important to the client's satisfaction with the provider and the process.

A knowledgeable representative of the integrator (if different from the system designer) should also be present to answer technical and site-specific questions. This person may be a senior technician, project manager or account representative for the project. The integrator usually is responsible for generating project-specific manuals or custom user guides which will be used during training. The trainer may also generate additional materials to support the training sessions.

Step Six: Sign-off and Start the Warranty Period

Even though commissioning and training are complete, more work must still be done. Contracts need to be closed out, warranty and preventive maintenance begins, and changes or additions to the systems, planned or unplanned, may be requested.

The Sign-Off

Upon completion of commissioning and training, it is time to sign off on the system. This is usually the responsibility of the owner who may be holding retainage as outlined in the contract with the integrator (see Chapter 3). This final approval may involve administrative layers, depending on whether or not the integrator is subcontracted to the GC or there is a construction manager who holds the construction contracts for the owner.

The basic requirements for final project sign-off are listed below:
- All punch list items are satisfactorily complete
- All documentation has been delivered
- System commissioning is complete
- Training needs have been met
- All contractual obligations have been met

Defining Warranties and Service Agreements

The project sign-off typically signals the beginning of the warranty period, though in some cases it may start before this point. In projects with a long timeframe, or those with delays caused by parties other than the integrator, the warranty period may begin at substantial completion[27] and/or when the client begins substantial use of the system (sometimes called "first beneficial use"). In a phased project, the warranty may begin in phases, timed to the completion of each room or area of the project. It is best to minimize the number of fragmented warranty dates to reduce confusion at the end of warranty periods.

---

[27] A sample substantial completion sign-off form is included in the appendix.

The term "warranty" requires clarification since there are three basic types of warranty, each with its own issues to address. The actual terms of the warranties can vary, depending on the individual manufacturers and the terms negotiated in a system-wide contract. The three types are:

- Manufacturer's warranty for the equipment only, which may sometimes have a service plan option
- System warranty, covering integration and integrator workmanship during the initial period following installation
- Service agreements, typically offered by the integrator to extend the base warranty period. Service agreements may also be entered into for an existing system not installed by the integrator, though this occurs less frequently.

Each of these is discussed in more detail below.

Manufacturer's Warranty

In most cases, manufacturer's warranties cover repair of specific equipment in an authorized service center. Removal and replacement of the equipment as installed, and transportation to/from a service center are not typically covered by the manufacturer. Additionally, a manufacturer's warranty may have differing duration periods for coverage of parts and labor.

The integrator may optionally, or by contract, extend the manufacturer's warranty to the installed location or to include parts and/or labor. Extension of a manufacturer's warranty terms by the integrator requires additional cost. If this coverage is contracted, the cost is included in the integrator's fee. This is a typical feature of a service agreement as described below.

A few manufacturers offer extended warranty plans including optional multi-year service plans. Typically, these extensions are at additional cost to the owner. If these plans are available, it is the responsibility of the integrator to make the owner and end-user team aware of them.

At sign-off, the integrator is responsible for delivery of documentation describing any manufacturer's warranties associated with equipment furnished by the integrator.

System Warranty

As a best practice (and typically required by contract), the integrator warrants all installation materials and workmanship against failure for a period of time after completion of the system installation, usually for a period of six to twelve months. This timeframe should be defined in the original project scope, specification or quotation. As noted previously, this typically includes an extension of any manufacturer's warranty that may be less than the initial system warranty timeframe.

While general terms of the warranty period should be well-defined in the contract, specific logistical terms need to be provided at sign-off. The integrator is responsible for providing a notice to the owner that defines the terms of any warranties and service agreements. This would include information and procedures for contacting the service provider's help desk or service department. A list of owner representatives who are authorized to call for service should also be provided.

Review the installation prior to the end of the warranty period or service contract

Toward the end of the warranty or service agreement period — usually four to six weeks before the term is to end — a representative of the integrator, owner, and any other appropriate stakeholders should review the installed system to ensure that all contractual obligations have been met and note any issues that may require resolution before the end of the contract term. The owner may choose at this time to negotiate the renewal of the contract and continue working with the integrator or use in-house staff, provided the adequate internal expertise and capacity have been developed.

Service Agreements

Once the system warranty is up, the owner may wish to contract for additional assistance and warranty coverage. The integrator may offer a service agreement contract that will provide continued warranty coverage and support beyond the initial period called for under the base system installation contract. The following components are normally included in this contract; each carries an associated cost and may be adjusted to meet the owner's needs.

- Extension of manufacturer's warranties including on-site service
- Warranty against system malfunction or failure
- Emergency and non-emergency service visits when required, including a maximum response time in the agreement
- Pre-planned preventive maintenance visits
- Pricing and terms for an optional extension of the contract

Managing Warranties and Service Agreements

For each type of warranty and service agreement, there are potentially terms and issues that are common to each. These generally relate to how the integrator reacts to the owner's and end-user's needs during this period. Response time and preventive maintenance are two of the primary issues, as explained below.

Response Time

Response time for service calls is negotiated per contract. A typical contract might include provision for a one-hour telephone response during business hours, same-day emergency response, and routine site visits by the end of the next business day. Some client sites are mission-critical and may require a twenty-four-hour, seven-day-per-week (24/7) response time. Response time requirements must be determined through coordination with the owner and end-users, and priced accordingly.

Preventive Maintenance

Preventive maintenance (PM) visits are strongly recommended. Depending on the usage and other site specifics, preventive maintenance might be performed monthly, quarterly or bi-annually. During each PM visit, each system and sub-system is nominally tested for proper operation, adjustments are checked and optimized, and system performance is reviewed with end-users. Following a PM visit, a report should be sent to the owner, describing the findings and adjustments made in the visit. Any recommendations for changes to the system or operational suggestions should be included.

PM visits should be scheduled around usage schedules and in keeping with the normal operation of the facility. If, during the design and pricing phase of the project, it is determined that PM must be performed outside of normal hours, an appropriate pricing structure should be in place. A PM visit is also a good time to offer follow-up training, ensuring that end-users are properly trained in use of the system.

A service record should be maintained. This record can be as simple as a three-ring binder kept in the equipment rack or a database system maintained by the service provider or owner. Whatever method is used, this record should contain end-user observations, services performed, and a record of particular equipment or system usage and downtime when this information is available. A service record that shows minimal downtime also shows that the service mission was properly planned and maintained.

Managing Changes to the System

There may be a need for modifications and adjustments to the system after it is commissioned and placed into service. Modifications include settings and alignments based on changing needs or preferences by users of the systems. Some changes may be significant enough to require an addition to the contract to implement.

Sometimes additions or changes may be required under the warranty or service agreement to maintain system functionality; these changes are typically covered under the service agreement. However, adjustments and modifications made by the owner, end-user or other party may void an existing warranty or service contract. To avoid breach of this contract, express written permission should be obtained from the service provider before such changes are made. Whenever changes are made to the system, any documentation associated with that change should be updated, including as-built drawings, equipment manuals and custom operating procedures. This is crucial to the ability to manage, update, troubleshoot and maintain the system over time, and, for many systems, is not an insignificant task.

Establishing a Long Term Professional Relationship

The key to a successful project is a system that a) meets the original design intent; b) fulfills the needs of the user; and c) remains in proper working order.

In many ways, the long-term success hinges on the ongoing professional relationship between the end-users and the AV service providers. In addition to responding to service calls while under contract, the provider should consciously develop this relationship by making periodic calls and contact to ensure the end-users have everything they need in their AV system.

The provider understands the nuances of the system's typical use and should be responsive to needs of the end-users. In return, the end-user should feel comfortable communicating routine observations and questions as they arise, enabling the service provider to respond in a timely manner. A service relationship built on this understanding helps avert unexpected system failures, minimizes emergency service calls and, most importantly, keeps the customer satisfied.

Be careful with software updates

The warranty and/or service agreement may (and should) include software and firmware updates. In general, these updates should only be performed when the update improves the performance or reliability of the system. Careful consideration should be made prior to the installation of any updates, including those to owner-furnished equipment (e.g., computers and network services). An update to one system or piece of equipment may trigger unexpected results from associated equipment or systems. In all cases, the integrator and/or service contract provider must be notified prior to update of any equipment associated with the system.

Chapter 7 Checklists

SUBSTANTIAL COMPLETION CHECKLIST

For the system to be substantially complete, almost everything must be installed and tested with preliminary record documentation available. This list contains the broad items that should be considered before declaring substantial completion and beginning the commissioning process.

Infrastructure and Related Systems:
- ❑ All AV rough-in and infrastructure in place
- ❑ Interfaces with other systems (lighting, drapes, screens, operable walls, etc.) complete and operational
- ❑ Communications services are installed and working (Ethernet, IP addressing, ISDN, telephone, and others)
- ❑ AV-related millwork and furnishings provided and integrated
- ❑ Operable walls for divisible rooms installed and adjusted
- ❑ Cable TV system installed and operational

Systems:
- ❑ All new AV equipment delivered and installed
- ❑ All AV OFE delivered and installed
- ❑ Initial control software uploaded, debugged and operation verified
- ❑ All DSP and other AV software programming and preliminary testing complete
- ❑ Preliminary testing and troubleshooting of signal paths for all video and audio systems
- ❑ Initial setup and alignment of video displays complete
- ❑ Initial setup and testing of all audio systems
- ❑ Programming and preliminary testing of conferencing systems
- ❑ All wireless devices tested and operational (microphones, control system devices, network devices)

General and Administrative:
- ❑ Major punch list items addressed
- ❑ Any known outstanding punch list items identified
- ❑ Marked-up shop drawings or preliminary as-built documentation on site

The final project documentation package should include the following.

As-built drawings, with final corrections, additions and field-verified information in the following formats:
- ❏ Full size paper
- ❏ Half size paper
- ❏ Electronic DWG
- ❏ Electronic PDF

- ❏ One-line system navigation drawing for larger or more complex projects

- ❏ Reduced as-built drawing sheets for use at equipment racks

Schedule & documentation of all physical system settings and adjustments including:
- ❏ Signal gain settings
- ❏ DSP and other software settings
- ❏ Codec settings
- ❏ Projector settings
- ❏ List of all static IP and telephone numbers
- ❏ Telephone and ISDN numbers
- ❏ All programming and equipment setting files on CD-ROM including current versions of manufacturer's editing and loading software.

- ❏ Test reports in paper and electronic formats.

- ❏ Manufacturers' users guides and manuals, alphabetized, bound in 3-ring binders with index. Include CD-ROM of any electronically available manuals.

- ❏ Excel spreadsheet of all equipment provided with all options and serial numbers noted.

System-specific custom operation guides including basic setup and operational procedures
- ❏ Laminated "Flash-card" style instructions for all basic system operations (e.g., playing a DVD, connecting a computer for display, volume and other audio settings, etc.).
- ❏ Basic troubleshooting guide in case of system malfunction, including common user errors and equipment failures

- ❏ List of consumable spare parts (lamps, filters, etc.)

- ❏ Key schedule with three duplicates of each key required for operation of the systems.

- ❏ Description of recommended service needs and intervals

Warranty statement, including
- ❏ System warranty start date, conditions, and term
- ❏ Summary of manufacturers' warranty coverage
- ❏ Description of extended warranties and service plans as purchased with the system

---

FINAL AV SYSTEM RECORD DOCUMENTATION PACKAGE CHECKLIST

At the end of the project, the items in this checklist are typically provided to the owner and end-users. The number of copies and the distribution should be determined in the contract. See the related Shop and As-Built Drawing Components Checklist for a more detailed list of the as-built drawing components.

CONCLUSION & AFTERWORD

CONCLUSION

Creating the best AV systems and environments requires an understanding of three fundamental aspects of AV integration: the technology, the human factor and the process. It is the process that ties the ever-changing technology and human factor constants together to create successful audiovisual systems and environments.

TECHNOLOGY CHANGES AND CHALLENGES AHEAD

It is through the extraordinary evolution of the AV industry in recent years that we have gained the knowledge documented in this book. Yet, while the future for AV and its place in the construction industry looks bright, the industry will continue to change, and new challenges lie ahead:

- As AV technology is absorbed into established IT networks, hardware and operations, what will be the impact on AV system design and integration (and their providers)?
- What solutions exist for AV system owners to manage their growing, changing and aging AV system inventories?
- What will it take to establish certification as a differentiator of excellence in service?
- When will technical performance standards for AV systems be fully established to support the needs of owners, AV designers and AV integrators?

These are questions that INFOCOMM INTERNATIONAL, members of the AV industry and others will continue to address in the years to come.

THE HUMAN FACTOR

In the meantime, the human factor — the driving force behind the technology — will remain unchanged. The need that humans have to clearly see images and hear the audio in rooms creates a common thread that will hold constant amidst the vast technological changes ahead for the AV industry. Creating good environments for AV will continue to be important for AV and construction industry professionals regardless of the technology behind the displays and the loudspeakers.

THE BEST PROCESS

Audiovisual Best Practices is a milestone for both the AV and construction industries. It was created to give all those involved in the AV design and integration process a common platform for communicating with one another about every facet and phase of the AV project. Following the steps in this guide, AV providers, building industry professionals, owners and end-users can now know what to do and what to expect in designing and integrating a system that best serves the end-users' needs.

AFTERWORD

THE FUTURE OF AV

If history is any indication, then the future bodes well for the pro-AV industry. Today, global networking, global partnerships and thus global communications – as facilitated by audiovisual systems – are prevalent.

In spite of apprehension that often accompanies change, the public has embraced the age of communication and is looking to new technology as a means to enhancing professional and personal lives. As tech-savvy children grow up to become managers, technicians, teachers and executives, they express an ever-increasing demand for technology to keep up with their expectations.

The pro-AV industry is up to the challenge. Its professionals have proven, since the early days of projection and audio recordings, that they understand consumer needs and can develop the necessary innovative products and value-added services to meet them.

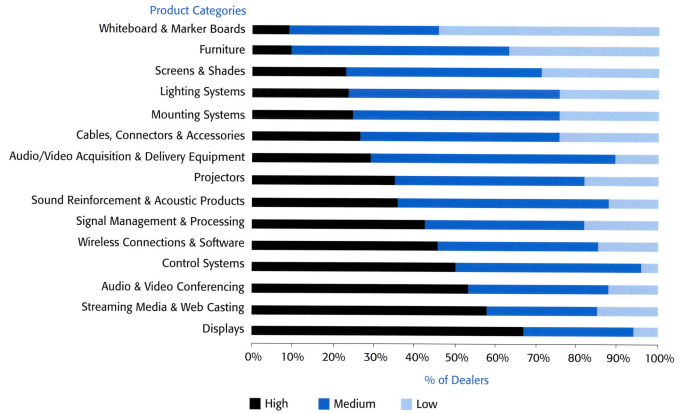

Figure 25. Degree of Opportunity for Revenue Growth by Product Category

According to the 2004 INFOCOMM INTERNATIONAL Market Definition and Strategy Study, two key factors are driving our growing market — new applications for AV products, and AV technology's accelerating convergence with IT. These trends are spearheading the spread of technologies like streaming media, webcasting, wireless technology, and AV-related software programs.

Source: INFOCOMM INTERNATIONAL 2004 AV Market Definition and Strategy Study

Audiovisual communication solutions abound in every aspect of daily life, bringing enhanced information to more people faster than ever before. Tied to networked systems and delivered via IT and telecommunications, AV has entered the mainstream of business, government, education, healthcare, worship, entertainment, broadcast and leisure activity.

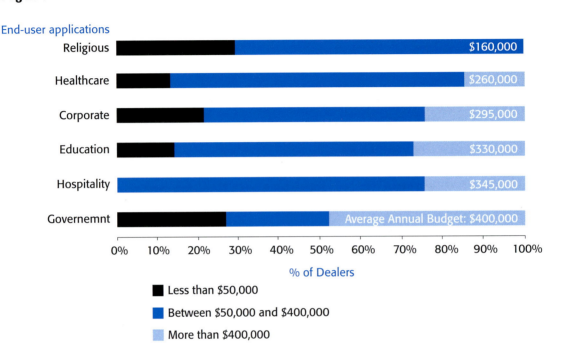

Figure 26. Most Attractive Markets Based on End-Users' Annual AV Expenditures

According to the 2004 Market Definition and Strategy Study conducted by InfoComm International, government, hospitality and education have the highest annual average budgets for AV equipment and services.

While AV solutions range from application to application, all promote better communication and understanding:

- An innovative solution at the Westchester Medical Center (NYC) helps handicapped pediatric patients escape into a beautifully colored virtual world without physical limitations.
- A distributed media system at Time Warner's headquarters helps visitors and staff stay connected and integrates the organization's vast portfolio of print, Internet, TV and film into a unified system for easy retrieval.
- Princeton Public Library installed the latest high end audio and video systems to enhance visitors' experience.
- At the Trinity Church in Manhattan, sound and video are key to enhanced delivery of message and music, encouraging congregational participation.

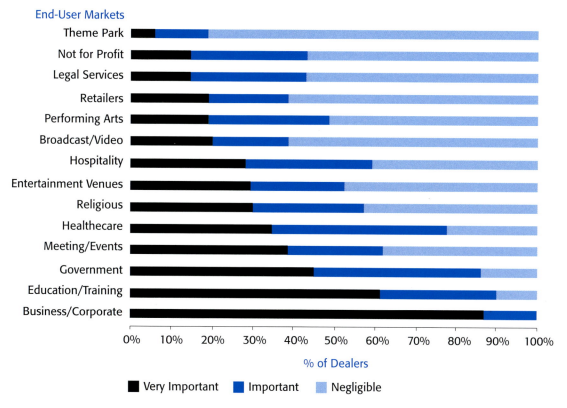

Figure 27. Dealer's Relative Assessment of the Most Important End-User Market Segments

The 2004 InfoComm International Market Definition and Strategy Study demonstrated which end-user markets appeared strongest in terms of audiovisual sales and services from the AV dealer perspective.

The list of recent innovations is extensive and diverse, including:
- The emergence of LCOS micro display technology as a competitor for plasma, LCD, and DLP
- A design for audio mixing consoles by an Australian company, which integrates audio console surfaces with an emphasis on ergonomic design
- A new product that allows the user to quickly isolate, identify, and resolve any interface problems that might interrupt a station's on-air capabilities
- Network-based, multi-screen display controller targeted for use in control rooms, communication hubs, network operation centers and briefing centers necessary to prepare government for the challenges ahead
- A technology that enables computer-generated information from anyone's laptop to be wirelessly transmitted to room displays
- Electronic signage networks (ESNs) once used only for commercial purposes, now being used for emergency alerts and warnings
- Digital voice-lift systems in elementary classrooms enabling all students to hear better, and therefore learn better
- Whole house automation systems that allow home owners to control household appliances, security systems and media devices remotely.
- Nanotechnology being brought to bear on electronics, promising better performance, lower cost and thinner form factors

APPENDICES

TABLE OF CONTENTS

Appendix I – Contract Forms and Checklists 181
    Sample Software License Agreement
    Standard AIA Project Forms and Templates
    The Brooks Act: Federal Government Selection of Architects and Engineers
    Owner Responsibility Checklist
    Statement of AV Integrator Qualifications

Appendix II – Program Phase 193
    AV Program Questions

Appendix III – Construction Phase 197
    Sample AV Systems Project Schedule

Appendix IV – Commissioning and Training 198
    Sample AV Substantial Completion Form
    Sample Warranty Terms
    Sample Custom User Guide
    Sample Custom Tech Manual

Appendix V – Resources and Bibliography 202

Appendix VI – Glossary 203

Appendix VII – Index 208

Appendix VIII – Authors' Biographies 214

I. Contract Forms and Checklists

SAMPLE SOFTWARE LICENSE AGREEMENT

This Software License is made by (Audiovisual Solutions Provider – AVSP), (address of AVSP), to the Customer as an essential element of the services to be rendered by (AVSP) as defined in the system specification and any associated documents and agreement. System shall mean the deliverable product as defined in these documents.

Customer and (AVSP) agree that this Software License is deemed to be part of, and subject to, the terms of the Agreement applicable to both parties.

SECTION 1   LICENSE GRANT AND OWNERSHIP

1.1   (AVSP) hereby grants to Customer a worldwide, perpetual, non-exclusive, non-transferable license to all software for Customer's use in connection with the establishment, use, maintenance and modification of the system implemented by (AVSP). Software shall mean executable object code of software programs and the patches, scripts, modifications, enhancements, designs, concepts or other materials that constitute the software programs necessary for the proper function and operation of the system as delivered by the (AVSP) and accepted by the Customer.

1.2   Except as expressly set forth in this paragraph, (AVSP) shall at all times own all intellectual property rights in the software. Any and all licenses, product warranties or service contracts provided by third parties in connection with any software, hardware or other software or services provided in the system shall be delivered to Customer for the sole benefit of Customer.

1.3   Customer may supply to (AVSP) or allow the (AVSP) to use certain proprietary information, including service marks, logos, graphics, software, documents and business information and plans that have been authored or pre-owned by Customer. All such intellectual property shall remain the exclusive property of Customer and shall not be used by (AVSP) for any purposes other than those associated with delivery of the system.

SECTION 2   COPIES, MODIFICATION, AND USE

2.1   Customer may make copies of the software for archival purposes and as required for modifications to the system. All copies and distribution of the software shall remain within the direct control of Customer and its representatives.

2.2   Customer may make modifications to the source code version of the software, if and only if the results of all such modifications are applied solely to the system. In no way does this Software License confer any right in Customer to license, sublicense, sell, or otherwise authorize the use of the software, whether in executable form, source code or otherwise, by any third parties, except in connection with the use of the system as part of Customer's business.

2.3   All express or implied warranties relating to the software shall be deemed null and void in case of any modification to the software made by any party other than (AVSP).

SAMPLE SOFTWARE LICENSE AGREEMENT

— Continued

SECTION 3 WARRANTIES AND REPRESENTATIONS

(AVSP) represents and warrants to Customer that:

3.1 it has all necessary rights and authority to execute and deliver this Software License and perform its obligations hereunder and to grant the rights granted under this Software License to Customer;

3.2 the goods and services provided by contractor under this Software License, including the software and all intellectual property provided hereunder, are original to (AVSP) or its subcontractors or partners; and

3.3 the software, as delivered as part of the system, will not infringe or otherwise violate the rights of any third party, or violate any applicable law, rule or regulation.

3.4 (AVSP) further represents and warrants that, throughout the System Warranty Period, the executable object code of software and the system will perform substantially in accordance with the System Specifications and Agreement. If the software fails to perform as specified and accepted all remedies are pursuant to the policies set forth in the Specification and in the Agreement. No warranty of any type or nature is provided for the source code version of the software which is delivered as is.

3.5 Except as expressly stated in this Agreement, there are no warranties, express or implied, including, but not limited to, the implied warranties of fitness for a particular purpose, of merchantability, or warranty of no infringement of third party intellectual property rights.

SECTION 4 INDEMNIFICATION

4.1 (AVSP) hereby indemnifies and shall defend and hold harmless Customer, its parent companies and its and their subsidiaries, affiliates, officers, directors, employees, agents and subcontractors from and against all liability, damages, loss, cost or expense, including but not limited to reasonable attorneys' fees and expenses, arising out of or in connection with any breach or alleged breach of the Agreement or any third party claims that the software or system here provided by (AVSP) infringes or otherwise violates any rights of any such third party.

4.2 Customer hereby indemnifies and shall defend and hold harmless (AVSP), its and their subsidiaries, affiliates, officers, directors, employees, agents and subcontractors from and against all liability, damages, loss, cost or expense, including but not limited to reasonable attorneys' fees and expenses, arising out of or in connection with any third party claims that Customer's use of the software in contravention of the grant of rights infringes or otherwise violates any rights of any such third party.

4.3 Upon the assertion of any claim or the commencement of any suit or proceeding against an indemnitee by any third party that may give rise to liability of an indemnitor hereunder, the indemnitee shall promptly notify the indemnitor of the existence of such a claim and shall give the indemnitor reasonable opportunity to defend and to settle the claim at its own expense and with counsel of its own selection. The indemnitee shall cooperate with the indemnitor, shall at all times have the right full to participate in such a defense at its own expense and shall not be obligated, against its consent, to participate in any settlement which it reasonably believes would have an adverse effect on its business.

SECTION 5   TRANSFER AND TERMINATION

This license will automatically terminate upon the disassembly of the system cited above, unless the system is reassembled in its original configuration in another location.

(AVSP) may terminate this license upon notice for failure to comply with any of terms set forth in this Software License. Upon termination, Customer is obligated to immediately destroy the software, including all copies and modifications.

As Customer's duly authorized representative, I have read and agree to this License.

Customer _____

Printed name of signatory _____

Signature _____ Date _____

Rev4. Sept 2003

Disclaimer: ICIA® provides this Sample Software License Agreement for illustration purposes only and makes no warranty, express or implied, including the warranties of merchantability and fitness for a particular purpose, or assumes any legal liability or responsibility for the accuracy, completeness, or usefulness of any information, or represents that its use would not infringe privately owned rights.

SAMPLE SOFTWARE LICENSE AGREEMENT

– Continued

STANDARD AIA PROJECT FORMS AND TEMPLATES

This is a partial list of forms available from the American Institute of Architects (AIA) either in electronic or paper form. Additional forms and templates are available. The forms are updated periodically, so check the AIA website for the most current information. Full descriptions of the forms are available at http://www.aia.org/docs_synopses.

Conventional Projects

A101™-1997	Standard Form of Agreement Between Owner and Contractor where the Basis of Payment is a Stipulated Sum
A107™-1997	Abbreviated Standard Form of Agreement Between Owner and Contractor for Construction Projects of Limited Scope Where the Basis of Payment is a Stipulated Sum
A111™-1997	Standard Form of Agreement Between Owner and Contractor Where the Basis of Payment is the Cost of the Work Plus a Fee with a Negotiated Guaranteed Maximum Price
A114™-2001	Standard Form of Agreement Between Owner and Contractor Where the Basis of Payment is the Cost of the Work Plus a Fee without a Guaranteed Maximum Price
A201™-1997	General Conditions of the Contract for Construction
A201™SC-1999	Federal Supplementary Conditions of the Contract for Construction
A401™-1997	Standard Form of Agreement Between Contractor and Subcontractor
A511™-1999	Guide for Supplementary Conditions
A701™-1997	Instructions to Bidders
B211™-2004	Standard Form of Architect's Services: Commissioning
C141™-1997	Standard Form of Agreement Between Architect and Consultant
C142™-1997	Abbreviated Standard Form of Agreement Between Architect and Consultant
C727™-1992	Standard Form of Agreement Between Architect and Consultant for Special Services

Small Projects

A105™-1993	Standard Form of Agreement Between Owner and Contractor for a Small Project, and
A205™-1993	General Conditions of Contract for Construction of a Small Project

Construction Manager Forms

A201™CMa-1992	General Conditions of the Contract for Construction, Construction Manager-Adviser Edition
A511™CMa-1993	Guide for Supplementary Conditions, Construction Manager-Adviser Edition
G701™CMa-1992	Change Order, Construction Manager-Adviser Edition
G702™CMa-1992	Application and Certificate for Payment, Construction Manager-Adviser Edition
G704™CMa-1992	Certificate of Substantial Completion, Construction Manager-Adviser Edition
G714™CMa-1992	Construction Change Directive, Construction Manager-Adviser Edition
G722™CMa-1992	Project Application and Project Certificate for Payment, Construction Manager-Adviser Edition

Interiors Forms

A175™ID-2003	Standard Form of Agreement Between Owner and Vendor for Furniture, Furnishings and Equipment where the basis of payment is a Stipulated Sum
A275™ID-2003	General Conditions of the Contract for Furniture, Furnishings, and Equipment
A775™ID-2003	Invitation and Instructions for Quotation for Furniture, Furnishings and Equipment

> **STANDARD AIA PROJECT FORMS AND TEMPLATES**
>
> – Continued

Design/Build Projects

A191™DB-1996	Standard Form of Agreement Between Owner and Design/Builder
A491™DB-1996	Standard Form of Agreement Between Design/Builder and Contractor
B901™DB-1996	Standard Form of Agreement Between Design/Builder and Architect
A141™-2004	Agreement Between Owner and Design-Builder
A142™-2004	Agreement Between Design-Builder and Contractor
B142™-2004	Agreement Between Owner and Consultant where the Owner Contemplates using the Design-Build Method of Project Delivery
G704/DB™-2004	Acknowledgement of Substantial Completion of a Design-Build Project

International Projects

B611™INT-2002	Standard Form of Agreement Between Client and Consultant for use where the Project is located outside the United States
B621™INT-2002	Abbreviated Standard Form of Agreement Between Client and Consultant for use where the Project is located outside the United States

Contract Administration and Project Management Forms

A305™-1986	Contractor's Qualification Statement
A310™-1970	Bid Bond
A312™-1984	Performance Bond and Payment Bond
B431™-1993	Architect's Qualification Statement
D200™-1995	Project Checklist
G607™-2000	Amendment to the Consultant Services Agreement
G612™-2001	Owner's Instructions to the Architect Regarding the Construction Contract, Insurance and Bonds, and Bidding Procedures
G701™-2001	Change Order
G702™-1992	Application and Certificate for Payment
G703™-1992	Application and Certificate for Payment Continuation Sheet
G704™-2000	Certificate of Substantial Completion
G706™-1994	Contractor's Affidavit of Payment of Debts and Claims
G706A™-1994	Contractor's Affidavit of Release of Liens
G707™-1994	Consent of Surety to Final Payment
G707A™-1994	Consent of Surety to Final Reduction in or Partial Release of Retainage
G709™-2001	Work Changes Proposal Request
G710™-1992	Architect's Supplemental Instructions
G711™-1972	Architect's Field Report
G712™-1972	Shop Drawing and Sample Record
G714™-2001	Construction Change Directive
G716™-2004	Request for Information
G805™-2001	List of Subcontractors
G806™-2001	Project Parameters Worksheet
G807™-2001	Project Team Directory
G809™-2001	Project Abstract
G810™-2001	Transmittal Letter

THE BROOKS ACT: FEDERAL GOVERNMENT SELECTION OF ARCHITECTS AND ENGINEERS

Public Law 92-582
92nd Congress, H.R. 12807
October 27, 1972

An Act

To amend the Federal Property and Administrative Services Act of 1949 in order to establish Federal policy concerning the selection of firms and individuals to perform architectural, engineering, and related services for the Federal Government.

Be it enacted by the Senate and House of Representatives of the United States of America in Congress assembled, That the Federal Property and Administrative Services Act of 1949 (40 U.S.C. 471 et seq.) is amended by adding at the end thereof the following new title:

"TITLE IX - SELECTION OF ARCHITECTS AND ENGINEERS "DEFINITIONS
"Sec.901. As used in this title

"(1) The term 'firm' means any individual, firm, partnership, corporation, association, or other legal entity permitted by law to practice the professions of architecture or engineering.

"(2) The term 'agency head' means the Secretary, Administrator, or head of a department, agency, or bureau of the Federal Government.

"(3) The term "architectural and engineering services" means -
1. professional services of an architectural or engineering nature, as defined by State law, if applicable, which are required to be performed or approved by a person licensed, registered, or certified to provide such services as described in this paragraph;
2. professional services of an architectural or engineering nature performed by contract that are associated with research, planning, development, design, construction, alteration, or repair of real property; and
3. such other professional services of an architectural or engineering nature, or incidental services, which members of the architectural and engineering professions (and individuals in their employ) may logically or justifiably perform, including studies, investigations, surveying and mapping, tests, evaluations, consultations, comprehensive planning, program management, conceptual designs, plans and specifications, value engineering, construction phase services, soils engineering, drawing reviews, preparation of operation and maintenance manuals, and other related services.

"POLICY
"Sec.902. The Congress hereby declares it to be the policy of the Federal Government to publicly announce all requirements for architectural and engineering services, and to negotiate contracts for architectural and engineering services on the basis of demonstrated competence and qualification for the type of professional services required and at fair and reasonable prices.

THE BROOKS ACT: FEDERAL GOVERNMENT SELECTION OF ARCHITECTS AND ENGINEERS

— Continued

"REQUESTS FOR DATA ON ARCHITECTURAL AND ENGINEERING SERVICES

"Sec.903. In the procurement of architectural and engineering services, the agency head shall encourage firms engaged in the lawful practice of their profession to submit annually a statement of qualifications and performance data. The agency head, for each proposed project, shall evaluate current statements of qualifications and performance data on file with the agency, together with those that may be submitted by other firms regarding the proposed project, and shall conduct discussions with no less than three firms regarding anticipated concepts and the relative utility of alternative methods of approach for furnishing the required services and then shall select therefrom, in order of preference, based upon criteria established and published by him, no less than three of the firms deemed to be the most highly qualified to provide the services required.

"NEGOTIATIONS OF CONTRACTS FOR ARCHITECTURAL AND ENGINEERING SERVICES

"Sec.904. (a) The agency head shall negotiate a contract with the highest qualified firm for architectural and engineering services at compensation which the agency head determines is fair and reasonable to the Government. In making such determination, the agency head shall take into account the estimated value of the services to be rendered, the scope, complexity, and professional nature thereof.

"(b) Should the agency head be unable to negotiate a satisfactory contract with the firm considered to be the most qualified, at a price he determines to be fair and reasonable to the Government, negotiations with that firm should be formally terminated. The agency head should then undertake negotiations with the second most qualified firm. Failing accord with the second most qualified firm, the agency head should terminate negotiations. The agency head should then undertake negotiations with the third most qualified firm.

"(c) Should the agency head be unable to negotiate a satisfactory contract with any of the selected firms, he shall select additional firms in order of their competence and qualification and continue negotiations in accordance with this section until an agreement is reached."

OWNER RESPONSIBILITY CHECKLIST

Throughout the process, the owner has a multitude of responsibilities as a participant in the overall project. This checklist provides some typical responsibilities that are specifically related to the AV work.

Programming
- Provide and coordinate appropriate owner personnel for participation in AV program interviews, meetings and benchmarking tours

Design
- Provide and coordinate appropriate owner personnel for participation in benchmarking tours and interviews
- Provide access to existing facilities for design team investigations and verifications of site conditions.
- Provide lists of any owner furnished AV equipment or related furniture
- Provide owner-specific contract requirements for bid package preparation

Bidding/RFP
- Provide access to existing sites for AV bid tours
- Provide appropriate personnel for any interview process for AV integrators

Construction

The owner may be responsible for the following items under any delivery method:

- Provide timely response and processing for project RFIs, change orders, pay requests and other AV-related communications
- Coordinate provision of any required central or offsite connections to HVAC and electrical services
- Meet OSHA requirements for the work environment for areas not controlled by the AV contractor
- Provide OFE equipment and furniture to AV integrator for installation
- Coordinate and negotiate software licensing agreement directly with the AV provider even if subcontracted through a GC
- Coordinate computer workstation installations in furniture with AV equipment installed
- Coordinate data network access for AV signal delivery if needed
- Coordinate the receipt of AV equipment and provide access and security during the installation process
- Coordinate server locations for video servers, help desk systems, control system servers and AV related web services where required
- Coordinate lighting, HVAC or other building system control with any building automation or energy control services in the final AV system installation
- Coordinate security of delivered AV equipment on site during period before contract sign-off and closeout

Depending on the project size, the contract terms and whether or not there is a general contractor on the job, the owner may be also responsible to provide the following items during construction under an AV design-build process. These should be specifically stated in the agreement with the integrator. Otherwise, these items would normally be provided by other contractors on a larger project instead of the owner directly.

- Provide conduit, backboxes and floor boxes for AV cabling and terminations
- Provide a dust-free environment for equipment delivery, storage and installation
- Provide cutting, patching and finish work related to base building changes required during AV installation
- Provide data/telecom cabling, outlets and service where required for AV use
- Provide AC power (which may require conditioning and/or isolated grounds) to AV device locations

OWNER RESPONSIBILITY CHECKLIST

— Continued

- Provide acoustics, noise and vibration control design and construction
- Provide low voltage interfaces to other new and existing peripheral systems such as lighting, projection screens, drapes, HVAC and other environmental devices
- Coordinate and/or provide millwork and furniture that will receive AV equipment if provided by the owner outside the AV contract
- Provide blocking and support for wall and ceiling-mounted AV racks and equipment
- Provide projection screen installation (The AV integrator may provide the screen but require that it be installed by others.)
- Provide required lighting design and installation for AV spaces

Testing, Commissioning and Training

- Provide and test communications services such as T1/E1, ISDN, telephone, satellite or other communications links that interface with the AV systems
- Coordinate LAN-based H.323 videoconferencing gatekeepers and gateways if used
- Coordinate IP addresses for any equipment on the LAN as needed
- Coordinate firewall configuration for LAN-based AV services if needed
- Facilitate schedules and owner personnel availability when connections to other existing AV systems are required
- Provide schedules of planned events and coordinate with commissioning schedule
- Coordinate and organize the end-users for system training
- Coordinate schedules and site access for the contractor when contracted directly to the owner

Appendix I — Contract Forms and Checklists

STATEMENT OF AV INTEGRATOR QUALIFICATIONS

Submitted To: _____
_____
_____

For Project: _____
_____
_____

Company Information:

Company Name _____
❏ Corporation   ❏ LLC   ❏ Partnership   ❏ Sole Proprietor   ❏ Joint Venture

Business Address (main office):
_____
_____

Telephone Number _____
Fax Number _____
Website _____
When Organized or Incorporated _____
Where Organized or Incorporated _____
Officers, General Partners and/or Owners _____

How many years have you been engaged in the AV contracting business under the present firm name? _____

Business Address (local project office, if applicable): _____
_____
_____

Telephone Number _____
Fax Number _____

Licensing, Bonding and Financial Information:

List current business and contractor licenses (including jurisdiction) held by the firm required for this project: _____
_____
_____

Is the Contractor's primary business (more than 50% gross revenue) as an electrical contractor?
_____

Does your company have an in-house AV installation department? _____
Credit available for new contracts $ _____
Gross Contracts currently in progress $ _____
Average yearly gross receipts for
    AV Integration projects in the past 5 years $ _____
Maximum bonding capacity $ _____
Bonding Company, Agent and Address   _____
                                                              _____
                                                              _____

Provide latest income statement and balance sheet (audited preferred) for the submitting organization, including name and address of preparer.

Provide a list of trade and bank references including contact information.

Relevant Experience:
List three representative projects completed during the last five years including the following information:
- Project Owner's name and address
- Name of primary contact and telephone number
- Date of final completion
- Design Consultant, if design-bid-build
- Initial contract amount
- Total of all change orders approved in addition to base contract
- Names of integrator's internal Project Engineer and Chief Installer who were responsible for each project.
- A brief description of the project. (Number of rooms, special capabilities, etc.)

(Provide space for requested information)

List Owner of the largest project completed during the last five years. Include:
- Project Owner's name and address
- Name of primary contact and telephone number.
- Date of final completion
- Design Consultant (if applicable)
- Contract amount and total of all Change Orders
- Total of all change orders approved in addition to base contract
- Names of integrator's internal Project Engineer and Chief Installer who were responsible for each project.
- A brief description of the project. (Number of rooms, special capabilities, etc.)

(Provide space for requested information)

List key personnel in your organization including their experience, certifications and their anticipated roles on this project. Also list any sub-contractors that you plan to use or regularly use and for what purpose.

Provide a summary of individual and company certifications. List each including the sponsoring organization, any specialty associated with a certification (e.g., PE-Electrical or CTS-I). For individual certifications, list each represented within the firm and the total number of individuals currently holding that certification. For company certifications, include the year that the current certification expires or requires renewal.

Legal Action and Claims:
Attach details for any questions below to which the answer is yes:

For the last three years, has your organization or its owners or officers been named in any contract-related litigation (including but not limited to litigation related to lien claims)? _____

In the last five years, has your organization or its owners or officers been involved in any bankruptcy or other insolvency proceedings? _____

STATEMENT OF AV INTEGRATOR QUALIFICATIONS

– Continued

STATEMENT OF AV INTEGRATOR QUALIFICATIONS

— Continued

In the last five years, has your organization failed to complete any awarded contracts? _____

In the last five years, have any owners or officers failed to complete any awarded contracts while with other organizations? _____

Has your organization or the undersigned ever been removed from, declared in default on, or been held in contempt of, a contract after being selected as the successful bidder? _____

Has your organization or the undersigned while at other firms ever had 5% (or more) of its total contract amount held as retainage for more than 12 months after substantial completion? _____

Insurance:
Workman's Compensation & Employers' Liability Insurance, Gross Amount: _____

Commercial Liability and other Insurance, Gross Amounts:
General Aggregate	$ _____
Products — Comp/OPS Aggregate	$ _____
Personal and Advertising Injury	$ _____
Each Occurrence	$ _____
Fire Damage (any one fire)	$ _____
Medical Expense (any one person)	$ _____

Signatures:
The Undersigned certifies under oath that the information provided herein is true and sufficiently complete so as not to be misleading:

Dated at _____

this _____ day of _____ 20 _____

Firm Name _____

By _____

Title _____

Notary:
_____ being duly sworn deposes and says that the information provided herein is true and sufficiently complete so as not to be misleading.

Subscribed and sworn before me this _____ day of _____ 20 ____

Notary Public:
My Commission Expires:

II. Program Phase

OVERVIEW, VISION AND FUNCTION

Project Overview and Vision
- Who are the key players to contact for obtaining information about the organization's vision for the future both organizationally and technologically?
- Is there currently a cohesive vision with regard to the future of technology in the organization?
- What are the primary goals for this facility/project?
- How will the AV and other technology systems proposed for this project support those goals?
- What is the overall application for the facility? (teaching? general conference? major convention? video origination? distance learning? performance? etc.)
- What level of sophistication is desired for the facility? state-of-the-art? middle-of-the-road? basic?
- From a technological standpoint, are there any unique features to the organization's operation?
- What functions and tasks are performed by the end-users of the facility?
- How does the facility plan to grow in the years to come?
- Are there specific allocations that need to be addressed for future expansion?
- What emerging technologies are considered by the owner and technology managers/end-users as desirable for the AV systems?

Standards and Benchmarks
- Are there known benchmark facilities which should be used as a model for this project from an architectural or technological standpoint?
- Are there any examples of technology inside or outside the organization that reflect the vision for technology?
- Are there any established standards for technology that need to be followed with respect to:
    - AV system design
    - Furniture design
    - Control system user interface design
    - Audio and video conferencing system design
    - Data/telecom structured cabling system design
    - Other systems
- Is documentation available for any of the standards?
- Are there any known challenges or deficiencies in the currently installed technology that need to be addressed?
- Are there any pilot projects or funding requests that may have an impact on the facility or systems designs?

Budget and Schedule
- Has a budget been established for technology systems?
- Are there any areas or systems which are of higher or lower priority with regard to budget?
- What is the base building schedule for design and construction phases?
- When is this facility scheduled to open?
- What is the current status of the project as a whole?

AV PROGRAM QUESTIONS

This is a collection of questions that can be used before and during AV program meetings to gather information about the owner, the end-users and their functional needs. The questions are grouped in three increasingly specific categories:

1. Overview, Vision and Function includes the vision questions and some overall information concerning benchmark standards, schedule and technical staffing. These are truly "program" level questions that require information from the users.

2. Global Technology Information includes more specific questions about technology that may be used building-wide, including cable TV and data/telecom infrastructure. This level of questions gets more specific about equipment and systems that may already exist, with some functional questions that are technically-oriented.

3. Function and Technology per Space includes questions that apply to each space being investigated. This level gets specific about functionality as well as the major components of the AV system required. Some technology-related questions may be answered by the AV provider based on the functional requirements, while others in this section require answers directly from the end-users or owner's technology managers.

AV PROGRAM QUESTIONS

— Continued

Technical Staff and Support
- Are the new systems going to be a significant change or addition to the existing technology in the organization?
- Is there an existing technical staff to manage, operate and maintain the existing AV systems?
- Are there specialized groups in the organization who address: AV systems? IT systems? distance learning systems? video production/post-production? training for end-users? others?
- Is there a current need for more technical staff?
- Will the current technical staff be sufficient for the new AV systems?
- Is there or will there be an AV help desk? How is it accessed by users?
- Is there a plan to outsource any technical support?
- Is space allocated yet for additional AV staff that may be required?
- Historically, what process has been used to design and install AV technology? In-house personnel? AV integrator design-build? AV consultant-led design-bid-build?

GLOBAL TECHNOLOGY INFORMATION

General Space and Technology Information
- What are the various types of spaces that require AV systems?
- Are any rooms to be dedicated theater-style seating with sloped floors and raised stage?
- Are tielines required between any of the spaces? If so, approximately how many and what type of tielines are required?
- Is there a need for a master control room to route signals between spaces or to other buildings?
- Where should equipment racks be located? Locally in each space? Any centralized equipment areas?
- Is there ample space allocated for the racks and any portable equipment which may need to be stored?
- How much portable equipment should be included in the base systems and how much will be rented or furnished by outside vendors?
- Is there existing AV equipment that is designated to be reused in the new facility?
- What documentation is available for any existing AV systems or spaces?
- How will ADA and Section 508 requirements affect the system?
- Is a paging system required? If so, what capability and priority does it need to have, and where shall the main station be located?
- Is there a fire alarm system with which the audio system needs to interface? How? mute? tone? voice announcement?
- Is a background music (BGM) system required? Is remote source selection or dayparting desired?
- Is a noise masking system required anywhere in the facility? If so, in what areas?

Cable TV, Electronic News Gathering (ENG), Satellite and Digital Signage
- Is there an existing cable or master antenna TV distribution system? If so, how many channels are currently offered?
- Where is the headend?
- What areas are currently served?
- Is there a need for a new or expanded TV distribution system throughout the building?
- Who will be responsible for the cable TV distribution system design?

- If so, from where should the feed be supplied?
- Where should new TV outlets (drops) be located?
- Will there be other in-building channels to be modulated (e.g., information channels? video bulletin boards? Video on-demand?)
- Is content from any AV spaces required to be delivered over the cable TV system?
- Is there any need for public area information displays or digital signage?
- Is satellite uplink or downlink capability available now?
- Where are the satellite dishes now? Where would new dishes go?
- Is there a need for an AV master control room for signal routing/distribution/media retrieval? video production?
- What type and how many tielines are required from master control to other spaces? other buildings?
- Should allocations be made for Electronic News Gathering (ENG) and Electronic Filed Production (EFP) connections for local and network news crews? If so, where?

Infrastructure/Data Network and Land-Based Communications
- What type(s) of connectivity does this facility require to other facilities or resources (fiber, T1, ISDN, cable TV, microwave, satellite up/down link, etc.)?
- What is the status of the data structured cabling system(s) for the facility? Are there any current projects already underway?
- Are there plans for upgrades?
- What type of data network is currently in place?
- Has a utilization study been conducted to evaluate existing network capacity?
- Have there been any efforts to date to deliver AV over the data network?
- Are there currently sufficient links between different facilities?
- What type and amount of fiber is currently installed?
- What fiber capacity is currently available for AV use?
- Is fiber currently used for the transport of AV signals?
- What documentation is available for existing structured cabling systems?

FUNCTION AND TECHNOLOGY PER SPACE
Space Specific Technology Program Questions
- Who are the primary users of this space?
- What are the functional goals of this space?
- What user tasks are associated with the space functions, and what AV technology applications are needed to support these tasks? AV-aided presentation? videoconferencing? computer training? internal meetings? education? government functions? external/community outreach?
- What are the percentages of use for each type of activity?
- How many people must be accommodated for each activity?
- Are there any specific, unique events/sessions planned for this space which would affect the AV system functionality?
- Are there any other known special audio and video requirements?
- What types of source material are being presented? computer graphics? text documents? powerpoint? spreadsheets? web pages? slides or film? computer code? art, fashion or architecture? movies? other entertainment? non-electronic documents or objects?
- Is multiple-image projection capability required for this space?
- Why? What images are being displayed simultaneously and for whom?

AV PROGRAM QUESTIONS

— Continued

AV PROGRAM QUESTIONS

— Continued

- For this space, should AV equipment be permanently installed or portable?
- Is the room divisible?
- If manual audio mixing is required, where will the mixer be located? Should it be portable?
- Is traditional slide projection capability required or is the image quality of slide-to-video converters acceptable? Is high-resolution slide-to-video conversion needed?
- Are there any tielines/inputs from other facilities or systems that are required for this space? cable TV feeds? central media feeds? Tielines to other spaces for overflow?
- Is videoconference/distance learning capability required for this space?
  If so, to what type of sites will this room be linked? What type of carrier is planned for videoconferencing? ISDN? T1? LAN/WAN?
- What type of video will be transported? NTSC/PAL/SECAM? HDTV? computer video?
- What is the bandwidth requirement for videoconferencing from this space?
- Is audio conferencing required with and separate from videoconferencing?
- Is there a need for enhanced video origination capability for this space (i.e., should 3-chip cameras, improved lighting, all sources gen-locked, and/or video effects equipment be incorporated into this design)?
- What audio and video recording capabilities are required for the space? What quality level is required? archival only? institutional? broadcast? Is a video server to be used?
- Is an Audience Response System required?
- Is ENG connectivity required for this space?
- What equipment is required for this space?

For major equipment and AV system requirements, identify the items below that may be required for this space. Enter a checkmark or the number of devices for items required for this room, Add "P" for portable, "C" for central media resource.

___ Ceiling-Mounted Video Projector(s) ___ Projector Lift(s)	___ Document Camera(s) ___ Traditional Overhead Projector(s)	___ VCR(s) ___ DVD Player(s) ___ CD Player(s) ___ Audio Cassette ___ Other _____
___ Fixed PC in Lectern/Desk ___ Laptop Connectivity at Lectern/Desk	___ Wireless Mouse & Keyboard ___ Electronic Marker Board	___ Slide Projector(s) ___ Slide-to-Video Converter
___ Stereo Audio Program Playback ___ Speech Reinforcement System	___ Cable TV/ Satellite Feed ___ Auxiliary AV Inputs	___ A/V Tielines for Taping Sessions & Overflow ___ Audio and Video Recording Capability
___ Touchscreen Panel(s) ___ Wall-mounted Controls ___ Audience Response System	AV Control Interface to: ___ Lights ___ Drapes ___ Screen ___ Other _____	___ Audio Conference Capability Videoconference Capability: ___ Single Axis ___ Dual Axis

III. Construction Phase

SAMPLE AV SYSTEM PROJECT SCHEDULE

Sample Work Breakdown Structure (WBS) in Gantt Chart View

ID	Task Name	Resources
1	1 Project Administration	
2	1.1 Notice of Project Award	sales
3	1.2 KICKOFF MEETING	all
4	1.3 Process Contract	admin
5	1.4 Pre Installation Project Planning	pm
6	2 Product Procurement	
7	2.1 Order Goods	purch
8	2.2 Receive Goods	warehse
9	3 System Design	
10	3.1 Team Review of Plans & Specs	all
11	3.2 Review of Millwork/Furniture Shop Drawings	eng
12	3.3 CAD – signal flows, rack elevations	cad
13	3.4 CAD – custom plates and assemblies	cad
14	3.5 CAD – cable pulls, risers, arch details	cad
15	3.6 Submit drawings for approval & coordination	pm
16	4 Software	
17	4.1 Review Design & Discovery w/ Client	prog
18	4.2 Design GUI	prog
19	4.3 GUI submittal/approval with end-user	prog
20	4.4 Write Code	prog
21	5 Fabrication	
22	5.1 Manufacture Custom Items	fab
23	5.2 Build & Wire Racks	fab
24	5.3 Shop QC & Software	prog
25	6 Site Installation	
26	6.1 Project Review w/ Site Techs	pm, eng
27	6.2 Cable Pulls & Rough-in	tech
28	6.3 Deliver Racks & Equipment to Site	warehse
29	6.4 Install/Terminate Racks & Equipment	tech
30	6.5 Prepare As Builts	tech, cad
31	6.6 Pre-Commissioning Tests	tech, qc
32	6.7 Report of Substantial Completion	pm
33	7 System Commissioning & Doc	
34	7.1 System Commissioning	qc/consult
35	7.2 Consultant Acceptance Test (if applicable)	consult
36	7.3 Custom and Manuf Manuals	doc
37	8 Project Close Out	
38	8.1 Deliver Final Documentation	pm
39	8.2 Training Session	pm
40	9 FINAL SIGNOFF	pm

The Project Manager's Tool: The Gantt Chart

The most common way to create a project schedule is by using a "Gantt chart," invented by Henry L. Gantt in 1917; this chart includes horizontal bars, which shows tasks and their relationship on a timeline. A Gantt chart is organized in logical groupings of tasks in a format similar to an outline. In traditional project management terms, this is referred to as a **Work Breakdown Structure (WBS)**. A WBS is a hierarchical notation that defines the tasks, activities, deliverables, and milestones required to complete a project. Each entry in the WBS has properties including, but not limited to, dependencies, budget, time duration, resource requirements, assignments and schedule restraints.

Of particular note in the construction schedule is the AV trade. AV cable pulls, structural rough-in, and other infrastructure should be accomplished during construction when possible and practical. Other than infrastructure, the bulk of an audiovisual system must be installed after all other trades are complete because of its sensitive nature.

IV. Commissioning and Training

SAMPLE AV SUBSTANTIAL COMPLETION FORM

At some point in the process, the system is declared substantially complete. This is often also when the system and/or equipment warranties begin. This is a sample acceptance form that would be signed by the owner to acknowledge substantial completion and begin the warranty. Depending on the contract arrangement, this form can refer to other documents such as the agreement originally signed with the owner or a warranty statement such as the sample form included in this Appendix.

Two variations on this form include a simple declaration of substantial completion for payment purposes without reference to the warranty or declaration of other project milestones (that may or may not include warranty start) such as final commissioning or training.

Audiovisual System Substantial Completion Form

[AV Integrator]

Project # _____
Date _____

Project Name: _____   PO # _____

Company Name: _____   Phone: _____

Address: _____

_____

_____

AV Sytsem Warranty Dates _____

Start _____

Installation Site: _____   End _____

Room Number(s): _____

Customer Contact: _____   Phone: _____

The audiovisual project described above has been found to be substantially complete and acceptable to the owner and/or their representative with the following exceptions:

1) _____

2) _____

3) _____

4) _____

5) _____

6) _____

7) _____

8) _____

9) _____

10) _____

By signature below, the owner acknowledges acceptance of the project and commencement of the warranty period for the systems noted according to the terms noted in the attached [warranty statement, contract, or specifications].

Name: _____   Title: _____

Signature: _____   Date: _____

Project: _____ Integrator's Project No.: _____
Location: _____ Customer Project No.: _____
_____ Warranty Start Date: _____

All new equipment contained in this system is warranted to be free of manufacturing defects per the terms and conditions of the original manufacturer's warranty. All manufacturers' warranties are honored and serviced by the AV Integrator.

A. DURATION

1. One (1) year, which will begin immediately following substantial completion of the installation, beneficial first use of the AV system or final acceptance of the system. The established start date is noted above.

THE WARRANTY SHALL INCLUDE:

1. On-site response within a maximum of ____ hours following receipt of the AV Integrator's standard Request for Service by an authorized representative.
2. Maximum ____ hour response for 24/7 telephone and/or fax and email support service for technical matters.
3. Parts and system components to restore system performance as required.
4. Labor to repair/service the system, components and parts in order to restore the system to complete operational condition.

THE WARRANTY SHALL NOT INCLUDE:

1. Replacement of consumable items such as video heads, optical blocks, plasma screens, LCD panels, CRT's, camera pick-up tubes or chips, projection tubes, lamps, batteries, and slide trays. Costs for such will be billed at current equipment and labor rates unless a manufacturing defect is discovered during the manufacturer's standard warranty period.
2. Requests for service not related to technical problems, but classified as "operator error".
3. Service required as a result of negligence, misuse, attempted repairs by anyone other than the AV integrator, or damage, or for equipment not related to the system supplied and installed by the AV Integrator.
4. Loaner equipment that is in place while the originally installed equipment it has replaced is under repair.
5. Connections to the contracted system made by others. *(Under these conditions service charges will be applicable as per our standard repair service policy.)*
6. Modifications to the system made by others, without prior written permission from the AV Integrator.
7. The cost to remove, reinstall and transportation to and from our service center, or the supplier/factory for components covered under their warranty.
   (Under these conditions, the warranty conditions will become null and void and charges will be applicable per our standard repair service policy.)

REQUESTS FOR SERVICE

Requests for service must be made by completion of the AV Integrator's standard request form. This outlines the conditions under which we will attend, and the costs for rectification of issues not covered under the warranty terms. This form must be received prior to dispatch of a repair technician.

REPORT

A written report/checklist will be issued following each repair/service, and must be acknowledged by an authorized representative.

SAMPLE WARRANTY TERMS

When the AV system installation is complete, there is normally a one-year warranty of the installed system and equipment. Similar terms can be purchased after this initial time period to extend equipment warranties and/or provide on-site service. Preventive maintenance site visits can be added to this type of contract and are recommended for any installed AV system where the owner has no internal AV technician team.

SAMPLE CUSTOM USER GUIDE

BOARDROOM SYSTEM   OPERATING INSTRUCTIONS • 9

Conducting an Audio Conference

This system can be used to audio conference with another location equipped for audio conferencing. You can transmit and receive local and far-end participant voices. Incoming audio projects from the ceiling loudspeakers.

Your conference will be most successful if you prepare the system in advance of connecting to the far-end location. To conduct an audio conference, you need to do the following tasks:

- Set up room environment.
- Connect to far-end location.

Procedure To prepare for and establish an audio conference, follow these steps:

1. On the touchpanel, touch **Audio Conference**.

Figure 3   Audio Conference Page

Setting up room environment

2. To set up the room environment, touch **Room Control** (refer to *Setting up the Room Environment* on page 13).

Connecting to far-end location

3. On touchpanel, use keypad to dial far-end location (dial 9 first). Numbers appear in display above keypad as they are touched. Touch **Backspace** to erase last number in display. Touch **Clear** to erase all numbers in display.

4. Touch **Connect**. You should hear dialing tones on local speakers. If line is busy, touch **Disconnect**. Wait a few minutes, and then touch Redial.

5. Upon successful connection, touch Volume arrows to adjust volume to a comfortable listening level.

6. Enjoy the conference.

Audiovisual Best Practices

SAMPLE CUSTOM TECH MANUAL

Equalizer Settings

The auditorium system includes three programmable equalizers: two for program audio and one for speech audio. The equalizers have been set up using the software included with your documentation package. The following files contain the scene settings:

- Speech EQ: SPEECH.SCN
- Program EQ: PROGRAM.SCN

CAUTION

The equalizers were set up by qualified technicians. Any changes to their programming might adversely affect sound quality.

Program Equalizer The program equalizer is configured as a 1/3-octave graphic equalizer whereby, the band filters act independently from each other, resulting in more prominent peaks and troughs in the frequency response. Figure 34 shows the device option settings.

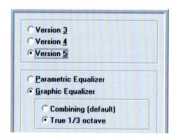

Figure 34   Program Equalizer Option Settings

Figure 35 illustrates the program equalizer settings and frequency response.

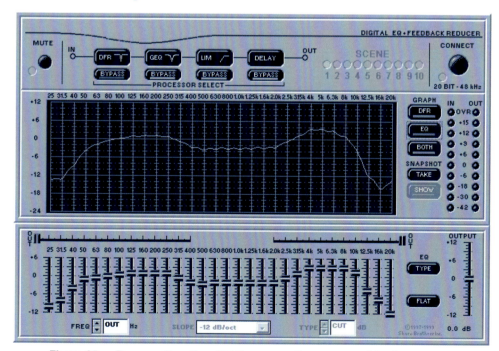

Figure 35   Program Equalizer EQ Frequency Response and Settings

Appendix IV — Commissioning and Training

V. Resources and Bibliography

Some general resources regarding the integration of AV and other technologies into buildings are available at the following websites:

- InfoComm International/InfoComm: http://www.infocomm.org
- AVolution: http://www.avolution.info
- BICSI http://www.bicsi.org

At this writing, the following websites offer information about the design and construction process and project management:

- The American Institute of Architects: http://www.aia.org
- A.S.A.P.M.: American Society for the Advancement of Project Management: http://www.asapm.org/
- CSI: The Construction Specifications Institute: http://www.csinet.org
- International Project Management Association (IPMA): http://www.ipma.ch
- Michigan QBS Coalition: Qualifications-Based Selection Awards; Selecting architects, engineers, land surveyors, design professionals - http://www.qbs-mi.org/faq.cfm
- Ohio State University Facilities Planning and Development Document Library: Project Delivery Manual: http://www.fpd.ohio-state.edu/?segue=MP
- Project Management Institute (PMI): http://www.pmi.org/
- Stanford University: The Project Delivery Process (PDP) at Stanford, Vol 1 and 2. http://cpm.stanford.edu/pdp.html
- The Royal Institute of British Architects: http://www.riba.org/

Updated resources and supplementary forms are available for download from the AV Best Practices website at:
www.infocomm.org/AVBestPractices

The following source is also available for specific information regarding the InfoComm International Dashboard for Controls project for AV control Graphical User Interfaces:
www.infocomm.org/dashboard

The following publications are recommended resources or were specifically referenced in *Audiovisual Best Practices: The Design and Integration Process for the AV and Construction Industries*:

- Construction Specifications Institute, **The Project Resource Manual**. Alexandria, VA: Construction Specifications Institute, 2004.
- Construction Specifications Institute, **MasterFormat 2004**. Alexandria, VA: Construction Specifications Institute, 2004.
- InfoComm International, Weems, CTS, MCP, Mike, ed. **Basics of Audio and Visual Systems Design**. Fairfax, Virginia: International Communications Industries Association, 2003.
- InfoComm International, **2004 Market Forecast Survey**, Fairfax, Virginia: International Communications Industries Association, 2004
- InfoComm International, **2005 Market Forecast Survey**, Fairfax, Virginia: International Communications Industries Association, 2004
- NAVA, Bolls, Terrence; Troup, Roger; McGee, Harry, **Audio-Visuals: A History**. Fairfax, Virginia: National Audio-Visual Association, 1980.

VI. Glossary

A

Acoustics: The scientific study of sound, especially of its generation, transmission, and reception.

A/E: Acronym for architect/engineer, usually referring to a firm that is primarily architectural with some in-house engineering services such as mechanical, electrical, plumbing and structural.

Applications: A specific activity or purpose that is supported by tasks. In the context of this book, training and teaching are applications that encompass tasks such as presenting information from a computer and taking notes.

Application Software: Computer programs or processes designed for performing specific tasks or uses.

Architect: As an individual, a registered design professional in the field of Architecture. As a firm, a team made up of such professionals, and typically the party responsible for the overall design and supervision of the construction of buildings or other large structures.

Audiovisual environment: The physical environment (usually the room and the building) in which any audiovisual systems will be installed or located.

AV: Acronym for audiovisual, sometimes also abbreviated as A/V.

AV cabling: Cables and wires for the transport of audio, video and audiovisual electronic signals.

AV control system: A collection of hardware and software designed and installed to allow users to control audio, video or audiovisual systems and devices.

AV manager: The person responsible for the management of a facility's or campus's audiovisual systems and technology, sometimes also known as technology manager.

B

Base building: Also known as "core and shell" or the "bricks-and-mortar," includes overall building elements such as structure, exterior walls and finishes, core (consisting of elevators, stairs, bathrooms, shafts and risers) and central mechanical and electrical systems; typically completed prior to tenant fit-up and often defined by the work in the General Contract.

Benchmarking: The process of examining methods, techniques, and principles from peer organizations and facilities, which is used as a basis for design of a new or renovated facility.

Best Practice: Methodology that, through experience and research, has proven to be the best path to a desired result. Recommended and agreed to by experts, best practices include the guidelines, processes, and procedures needed to accomplish an objective effectively, efficiently, and with beneficial results.

Bidding process: A procurement process in which proposals are solicited from contractor(s) or service provider(s) for the purchase of goods, services, or finished systems and buildings. Bidders' proposals may include qualifications, pricing, clarification of methods, or value engineering and are typically based on terms and specifications provided to bidders at the time of solicitation.

Bids: Offers or proposals to supply specific services and/or equipment under the terms and for the price identified in the bids.

Bid bond: A surety bond often required of bidders on construction work to ensure that the successful bidder will accept the job. See Performance Bond.

Box sales: Sale of equipment or accessories without design or installation services.

Box sale vendor or dealer: A firm or party offering box sales. An AV dealer.

BRI: acronym for Basic Rate Interface; a single ISDN line that utilizes two 56 or 64Kbps channels plus one 16Kbps data channels. *see also* PRI and ISDN.

Business model: The means and methods by which a company generates revenue and profit.

C

CA: Acronym for Construction Administration, which includes the activities of the design team and/or the owner during the construction phase that pertain to the administration of the construction contracts with the GC and subcontractors.

CAD: Acronym for Computer Aided Drafting, an electronic method of creating technical drawings. Also referred to as CADD, Computer Aided Drafting and Design.

Capital funds: Funds for purchase or improvements of buildings and equipment. These are long-term funds handled separately from short-term operational funds.

Casework: Architecturally integrated cases and cabinets, usually based on wood products for construction. See millwork.

CD: Acronym for Construction Documents, which includes the finalization of the drawings and specifications in preparation for contracting and construction. The construction documents are sometimes referred to as CDs. CD can also refer to the complete set of documents including drawings, specifications and contract agreements known as the Contract Documents.

Certification: A process, often voluntary, by which acknowledgement is accorded to individuals who demonstrate a level of knowledge and skill required in a profession or product. Certification is typically qualified and issued by a trade organization or manufacturer. Certification offered by publisher, InfoComm International, includes Certified Technology Specialist (CTS) designation recognizing general, design or installation proficiency, and the company level Certified Audiovisual Solutions Provider (CAVSP) designation. Also see License.

Change Order: Often abbreviated 'CO'. A written document regarding a change to the work after a bid is awarded and a contract is executed. Once approved by the appropriate parties, this document amends or "changes" the contract drawings, specifications, scope of work and/or pricing.

Command and Control Center: Centralized operations and/or dispatch center for management of large-scale systems. Also known as Network Control/Operation Centers (NCC or NOC). Used by network/Internet service providers, suppliers of electricity or natural gas, building management teams, and other organizations that run large systems that require centralized monitoring and control.

Commissioning: The test and alignment process to verify that a system is designed, installed, adjusted, and functioning optimally according to the design intent and contract requirements.

Conduit: A tube or duct for enclosing wires or cables; may be metallic or non-metallic.

Construction manager: A construction manager may be an individual or company that is hired to manage the design and construction process as a representative of the owner. Sometimes also called the Program Manager.

Consultant: In the construction industry, the designers of systems who contract with either the architect or owner.

Contract fees: Fees for services or equipment defined in a legally binding contract.

CTS: Acronym for Certified Technology Specialist, a certification offered by InfoComm International. Several specialty CTS certifications are available, including CTS-D (Design) and CTS-I (Installation). For more information see the Introduction or visit www.infocomm.org.

D

Data/telecom: General term referring to data networking, telephone and other digital and analog communications technologies and systems.

DD: Acronym for Design Development, which includes the activities of the design team after conceptual and schematic design and leading into the construction documents phase.

Dealer: An authorized representative and reseller of manufacturers' equipment; see Integrator.

Delivery options: Options in how the work is procured.

Delivery process: The overall process for delivery of a project from concept through completion.

Design-Bid-Build: A building or system delivery method under which design and construction or integration are provided under separate contracts, with design services usually provided by an independent consultant. See Chapter 2.

Design-Build: A building or system delivery method under which both design and construction or integration under a single contract. See Chapter 2 for more information.

Distribution systems: Systems used to distribute signals such as audio, video, data and control to multiple endpoints.

E

E/A: Acronym for engineer/architect, usually referring to a consulting engineering firm that also offers architectural services.

EC: Acronym for Electrical Contractor, the party typically responsible for the supply and construction of the electrical systems and components in a construction project.

Electrical engineer: A registered professional engineer in the field of Electrical Engineering; the party typically responsible for the design of the electrical systems in a construction project.

End-user: The person or group who uses the AV systems or equipment for their intended purpose after installation and commissioning are complete.

Equipment sales incentive: see Sales incentive.

F

Facility manager: The person responsible for the overall operations and maintenance of a facility as an employee of or contracted by the facility owner.

Fast-track: A design and construction process intended to expedite a project by altering and shortening the sequence of the design and construction, often including having traditionally sequential elements occurring in parallel.

FF&E: Acronym. See Furniture, fixtures and equipment.

Finishes: The treatment of visible surfaces in a room such as wall surfaces, floor surfaces, ceiling surfaces and furniture surfaces. Finishes include paint, laminates, fabrics, acoustical paneling, ceiling tiles, carpet, glass, and other finish materials.

Furniture, fixtures and equipment (FF&E): Items not normally considered permanently attached to a structure, but which are counted as a bondable cost in new construction or major renovation projects. These items are often funded separately from the base building capital funding.

G

Gantt chart: Used as a project management scheduling tool, with a horizontal bar graph to show plan or progress for each task within a project. Referenced against a horizontal time scale, each task's bar shows its start date, duration, and end date. A Gantt chart illustrates the established Work Breakdown Structure (WBS) of the project.

GC: Acronym. See General Contractor.

General Contractor (GC): A person or business entity that is contracted to be in charge of a building project construction team usually involving the use of subcontractors. Also called lead contractor or prime contractor. Most states require licensing of general contractors.

H

High voltage: Per the National Electrical Code (NEC), circuits and systems operating at over 600V; per the IEEE circuits and systems operating at over 1000V; commonly used in the AV industry to reference systems operating at 110V or greater.

HVAC: Acronym for Heating, Ventilation and Air Conditioning, a range of engineering and construction services often offered by a single entity.

I

ICAT: Independent Consultants in Audiovisual Technology Council, one of the member councils of InfoComm International, formed as a peer group for independent consultants. Along with the SAVVI and Technology Managers/End-Users Councils, ICAT was instrumental in advancing the publication of this guidebook.

ICIA®: Acronym for the International Communications Industries Association, Inc®; the trade association representing the professional audiovisual industry worldwide, also associated with its annual tradeshow, InfoComm. More information is available at www.infocomm.org.

Independent consultant: An individual or firm providing professional consulting and/or design services, which has no direct financial relationship with or obligation to any manufacturers, representatives, dealers, contractors or integrators.

InfoComm: See InfoComm International.

Infrastructure: In terms of an AV project, the basic facility services and installations required to support the functioning of an installed audiovisual system, including conduit, power, structural supports, cooling, data/telecom services, space planning and other aspects of base building that affect AV system operation and use.

In-house: Work or services performed or provided within a single firm without having to contract outside the firm.

Installation: The physical act of the construction of the work.

Installer: A person or firm providing installation services.

Integrator: An authorized representative and reseller of manufacturers' equipment that also offers design, installation, and repair services.

Integrator-Led Design-Build: A project process in which an AV integrator is contracted for both design and installation services. These services may be contracted directly to the owner or separately to the architect and general contractor. See Chapter 2.

ISDN: Acronym for Integrated Services Digital Network; a switched data network providing end-to-end connectivity, most notably for use for telephony and videoconferencing. see also PRI and BRI.

Isolated Ground (IG): Electrical system whereby the equipment ground is isolated from all building structure and conduit, and is not shared between multiple devices or branch circuits. IG wiring helps protect electronic equipment from noise generated and/or transmitted through the electrical system and requires special receptacles and wiring. Specified in NEC, Sec. 250-74.

ISP: Acronym for Internet Service Provider.

IT: Acronym for Information Technology; in corporate departmental terms, also referred to as Information Services (IS). (Sometimes this acronym is also used to represent Instructional Technology in some educational settings.)

L

LAN: Acronym for Local Area Network; a term used to denote the data network that serves an area, department, floor or building.

License: A legal credential obtained by an individual or business, usually issued by a government agency, which is required to perform work in a particular trade and/or jurisdiction. A license is often required by law or ordinance for workers in various building trades, including architecture, mechanical, electrical, and structural engineering, some low-voltage trades and other building design, engineering and construction roles.

Line item: A specific service or piece of equipment identified as a separate item in a price list, bid or specification.

Low voltage: Per National Electrical Code (NEC) ARTICLE 720, circuits and equipment operating at less than 50 Volts, however often commonly used by NEC to also reference Power Limited Circuits defined as Class 2 (less than 30V and less than 100VA) and Class 3 (greater than 30V and less than 100VA) circuits. Often also used to refer to the building-related electronics trades such as AV, data/telecom, fire protection and others.

M

Mark-up: An amount added to a cost price in calculating a selling price, especially an amount that takes into account overhead and profit.

Mechanical Engineer: A registered professional engineer in the field of Mechanical Engineering; the party typically responsible for the design of the HVAC systems in a construction project.

MEP engineering: Mechanical/Electrical/Plumbing engineering.

Millwork: Woodwork, such as doors, window casings, and baseboards, ready-made by a lumber mill or finished work by a woodworker.

N

NEC: Acronym for National Electric Code. Electrical safety code adopted in the United States; the standard on which most electrical and fire-related laws and ordinances are based.. Developed by the American National Standards Institute (ANSI) and sponsored by the National Fire Protection Association (NFPA). See http://www.nfpa.org/.

Needs analysis: The process used to establish the needs or requirements of a project or system including functional, operational and budgetary requirements; in the architectural, design and building industries, this process is known as the Program Phase. See Program.

O

OFE: Acronym for Owner Furnished Equipment; also known as CFE (Customer Furnished Equipment).

Off-site: Most often used to refer to a location other than the project construction site.

Outsourcing: The procuring of services or products from an outside supplier or manufacturer, in lieu of using or developing in-house capabilities.

P

Parameters: Measurements or values on which something else depends.

PCO: Acronym. See Proposed Change Order.

PE: acronym: see Professional Engineer.

Performance bond: A surety bond issued to one party of a contract as a guarantee against the failure of the other party to meet obligations specified in the contract.

Plenum: The air-handling space found above ceilings or below raised floors; most commonly the return-air path for an HVAC system. Federal and local laws require fire-code ratings for any materials installed or located within this space.

POTS: Acronym for Plain Old Telephone System or Service. This refers to the traditional analog telephone line.

Prime contract: The contract for the party having overall responsibility for the services or work referenced in the contract.

Pro-AV, or pro AV: Short form for professional audiovisual. This refers to products and systems that incorporate commercial-grade components for mostly commercial, government and educational facilities. These products and systems are differentiated from consumer and residential AV.

PRI: Acronym for Primary Rate Interface; a multi-channel ISDN line consisting of twenty-three 56 or 64 kbps 'B' (data) channels and one 16 kbps 'D' (signaling) channel. Similar in bandwidth to an E-1 or T-1 line. see also ISDN and BRI.

Procurement: The process associated with the acquisition and/or purchase of goods and services.

Professional Engineer: A registered and/or licensed engineer, often associated with design building disciplines such as electrical, mechanical, plumbing, structural and others. See the National Society for Professional Engineers website http://www.nspe.org/.

Program: A document summarizing the results of a needs analysis. During the program phase, the architect, AV professional and other design team members discover the end-user's needs by examining the required application(s), the tasks and functions that support the application, and the wishes and desires of the end-user. (Alternate meaning: software program. See Application Software.)

Program manager: The program manager's job is to make sure the owner's needs are met according to the architectural and system program that has been established. This function is sometimes combined with the Construction Manager role.

Programming: In the electronics trades, generally refers to writing code to operate an electronic device, e.g. a control system or DSP processor. (Alternate meaning: the act of developing the architectural and AV program or needs analysis.)

Project Manager (PM): The individual (or sometimes a firm) who oversees, plans, and coordinates a project, or delineated part of a project. The PM is responsible for a team's budget, schedule, administration, and adherence to contract.

Project model: The process selected which defines the means and methods to be applied to a project for design and procurement.

Project schedule: A schedule developed specifically for a project identifying coordination with other trades and any relevant milestones or deadlines. A project schedule is often documented through the use of project management tools such as a Gantt chart and Work Breakdown Structure (WBS).

Project type: A delineation of a project based on descriptors such as the project model or delivery method, the vertical market in which the project is being built and the type of systems involved.

Proposed Change Order (PCO): A Change Order that has been proposed but not yet approved or accepted by the appropriate parties. Similar to Request for Change; see Change Order.

Q

Quality control/assurance: A system for ensuring the maintenance of proper standards and quality of manufactured goods or systems, especially by periodic random inspection of the product

R

Renovation: Usually refers to the demolition, re-design and re-building of a space or building, often with modifications to the floor plan and possibly adapted for re-use to the same or a different purpose. This is sometimes distinguished from restoration.

Restoration: Usually refers to a specific type of renovation of an existing space or building with the intent of retaining its existing purpose and/or restoring the space to it's original look and use.

Retrofit: The replacement of existing devices or systems with newer items or systems using the existing infrastructure or within the existing space.

RFI: Acronym for Request For Information. A project document submitted during bidding or construction to obtain additional information or clarification.

S

Sales incentive: A financial incentive related directly to the sale of equipment or services, typically offered to sales personnel as incentive to meet quotas. Equipment manufacturers often establish Sales Promotion Incentive Funds (SPIFs) to support these incentives.

SAVVI: Sound, Audiovisual and Video Integrators Council; one of the member councils of InfoComm International, formed as a peer group for AV integrators and dealers. Along with the ICAT and Technology Managers/End-Users Councils, SAVVI was instrumental in advancing the publication of this guidebook.

Shop drawings: Set of drawings that contain all details required to fabricate and install a system per the contract intent. Details include equipment, wiring, mounting, plans, elevations, sightlines, wire-tags, pin-outs, etc.

"Shopping" the design/proposal: The distribution and use of one provider's proposal or design to solicit similar proposals from other parties with the intent of obtaining the lowest price. When done without the prior consent of the original provider, the practice is often considered unethical.

Site monitoring: Oversight at the project site.

SPIF: Acronym for Sales Promotion Incentive. See Sales incentive.

Stand-alone upgrade: An equipment or system upgrade not affecting other systems or devices.

Start-up: Initial activation and operation of a system.

Structural: Related to the physical structure of a building and often requiring safety considerations for support and mounting of equipment and devices.

Structural engineer: A registered Professional Engineer in the field of Structural Engineering; the party typically responsible for the design of the structural systems in a construction project.

Subcontractor: A firm or individual providing services under a contractual agreement to another contractor.

Submittal: In construction, submittal refers to the documents, samples and information to be submitted to the project designers, the owner or others in a defined manner normally identified in the Contract Documents.

Surety bond: In the construction industry, this most often refers to an agreement or insurance policy that provides monetary compensation to the owner should a contractor fail to perform. See bid bond and performance bond.

System concept: A conceptual-level design of the system indicating general requirements, primary devices and major subsystems.

System requirements: See Needs analysis.

System upgrade: A functional or performance enhancement to a system.

T

Technical representative: A representative, usually an Owner or End User, familiar with the technology and/or technical requirements of a project.

Technical support: Operations and/or physical support and maintenance of the systems and devices.

Technology-rich: Enhanced technical systems or capabilities.

Third-party: Any person or firm who is not a party to a contract. This term is often used to refer to someone who may become involved in or affected by a contractual relationship between others.

T-1/E-1: A dedicated high-speed digital connection commonly used to connect businesses to telephone and data networks. The North American T-1 supports data at rates up to 1.544 megabits per second on 24 data channels (also referred to as DS-1, or Digital Service - Level 1); the European E-1 supports up to 2 mbps on 32 data channels.

Timelines: A schedule identifying specific deadlines or milestones. In the construction industry, they are often expressed in a Gantt chart.

Training: Education of the designated Owner's representative, technology managers, and/or end-users for the operations, maintenance, and support of a system or device.

Turnkey: A complete system solution; supplied, installed or purchased in a condition ready for immediate use, occupation or operation in the audiovisual industry, normally applied to a "one-stop" process with a single contracted entity responsible for all facets of a project process, from Program through Commissioning.

U

UTP: Acronym for unshielded twisted pair. Used to generically denote cabling that includes two insulated copper conductors twisted together without any shielding conductor surrounding the twisted pair. Common telephone cabling and data cabling (including Category 3, 5, 5e and 7) as well as typical audio speaker cabling is considered to be UTP.

V

Value engineering: An organized effort directed at analyzing the function of the systems, equipment, facilities, procedures, and supplies for the purpose of achieving the required function at the lowest total cost of effective ownership, consistent with requirements for performance, reliability, quality, and maintainability. This should not be confused with budget or cost reduction that simply reduces the scope of work or negatively affects the project's performance, reliability, quality or maintainability.

Vertical market: Market segment, delineated by market type or industry such as religious, corporate, educational, government, medical, etc.

Videoconferencing: Communication between two or more physically remote parties with picture and sound, normally with real-time or near real-time interaction.

W

Warranty: An assurance by the seller of goods or services that the goods or services are as represented or will be as promised; guaranties given to the purchaser by a company stating that a product or system is reliable and free from known defects and that the seller will, without charge, repair or replace defective parts within a given time limit and under certain conditions.

Work Breakdown Structure (WBS): Hierarchical listing of tasks and sub-tasks required for completion of a project, organized by trade, team, function or phase. Often represented in the form of a Gantt chart, a WBS is articulated so as to represent dependencies between tasks, critical paths, and resource needs. A sample WBS is included in the Appendix.

Work directive: A written directive relating to the work being performed.

VII. Index

A/E (Architect/Engineer), 27, 69, 203
Acoustical consultant, 25, 30, 104
Acoustics, 6, 10, 18, 92
    classroom acoustic standard, 184
    consulting, 12, 28, 47, 48, 51, 54
    coordination, 45, 106, 109
    defined, 205
    in checklists, 76, 102, 126, 189
Added value, 12
Alternates, 61, 112, 117, 127
Americans with Disabilities Act (ADA), 85, 101, 194
Applications, 80-82, 91
    defined, 203
    for payment 114, 184-185
    in benchmarking, 110-111
    in checklist, 193
Architect, 7, 33-37
    and AV providers, 57, 63, 65
    and contracts, 46-54, 67, 69, 112, 121
    and coordination, 30-31, 34, 131
    and design team 25, 27-28, 56
    and program report 98
    defined, 203
    in checklists, 184
    in conceptual design, 21
    in construction phase, 135, 138-139
    in design phase, 104, 106, 107
As-builts, 137-139, 143, 146, 160, 169
    in checklists, 77, 170-171
Audioconferencing, *see Conferencing*
Audiovisual, *see AV*
Audiovisual Solutions Provider (AVSP) 11, 15-17, *see also CAVSP*
    in forms, 181-182
Automation
    building, 28
    home, 3, 9, 177
    in checklists, 100, 188
AV
    and documentation, 139, 146, 160, 165
    and warranties, 105, 167
    best practices, *see Best practices*
    consultant, *see Consultant, independent AV design,* control system, *see Control systems*
    dealer, 9, 10, 12, 53, 59, 121, 177, 203
    department, 11, 26-27
    designer, 22, 25, 27-30, 35, 56,
    see also consultant, independent AV
        and budgeting, 122
        construction phase, 139, 157, 162, 163, 174
        contracts, 68, 74
        design phase, 105, 107, 109, 116, 123
        program phase, 88-89, 93, 99
    distributor, 12
    environment, 6, 8, 108, 132-133
        in checklists, 150, 188
    future of, 5, 174-175
    history of, 2, 5, 202
    in checklists, 151, 171, 199
    industry, 2-18, 24, 43, 174-178
    installation, 21, 31, 105, 130-131
    integrator, *see Integrator*
    manager, 26-27, 94, 203
    manufacturers, 12, 111, 118, 140-141, 144, *see also AV integrator*
    market segments, 3, 8, 9 177, 207
    process, 20-21. 44-45, 83, 105, 137, 140, 147
    provider, *see Consultant, Dealer, Integrator, AV resources*
    resources, 12, 15-17, 202
    solutions provider, *see Audiovisual Solutions Provider (AVSP)*
    tasks and parameters, 90, 100
    technology manager, *see AV manager*
AV systems,
    functionality, 80-82, 89, 92, 100, 110, 116, 120, 128, 143-144. 154, 157, 162, 193
    funding, 42-43, 97, 108-109, 163
    integrator, *see Integrator*
AVSP, *see Audiovisual Solutions Provider*

Base building, 20-22, 27, 44-45, 109, 112, 126, 130, 140, 203
Basics of Audio and Video Systems Design, iii, 202
Benchmarking
    defined, 203
    design phase, 110
    program phase, 86-87, 102, 193
Best practices, 174, 202
    AV manager, 26
    budgeting, 34, 42
    cabling contracts, 72
    consultants and designers, 28-29
    contracts, 68, 70-71
    defined, 203
    design reviews, 123, 124
    education, 43
    GUI design, 122, 145
    infrastructure design, 107
    owner, 27
    owner move-in, 133
    procuring AV equipment, 140
    program phase, 98
    quality, time and money 111
    selecting AV team, 116
    site security, 146
    software updates, 169
    submittal reviews, 139
    team work, 132
    training,165-166
    warranties, 167
BICSI (Building Industry Consulting Service International), 15, 202
Bids and Bidding, 22, 42, 77
    addenda, 66
    Internet, 60
    invitation to, 112, 127, 184
    methods, 57-60
    middle bid method, 59-60
    package, 63
    process, 65-66
    qualifications, *see Request for qualifications (RFQ)*
    response form, 112, 127
Bonding 62, 78, 113, 135, 148, 185, 203
Box sales, 203, *see AV dealer*
Brooks Act, 58, 186-187
Budget, 28, 34, 42, 50, 76, 95-96, 101, 106, 108-110, 149, 176, 193, *see also AV systems: funding*
    line item, defined, 205
    management, 122-123
    use in delivery method selection, 50
Building committee, 27
Building management agency, 34
Building Industry Consulting Service International, *see BICSI,* 24
Business model, 203
Buyer, *see Purchasing Agent*

CA, *see Construction administration*
Cable TV, 30, 133, 162
    in checklists, 148, 150, 170, 193-196
Cabling, 8, 30, 72, 130-132, 141-142, 197, *see also structured cabling, data/telecom*
    pathway, 28, 107
    tag reference, 142
CAD, CADD (Computer Aided Drafting and Design), 138, 203
CAVSP, *see Certified Audiovisual Solutions Provider*
Casework, 32, 130, *see also Millwork*
    defined, 203
    in checklists, 148
CD, *see Construction documents*
Certification, 13-15, 35, 116, 174, 204, 214, *see also Certified Technology Specialist, Certified Audiovisual Solutions Provider*
    defined, 203
    in checklists, 78, 191
Certified Audiovisual Solutions Provider (CAVSP) 13, 116, 203, *see also AVSP*
Certified Technology Specialist (CTS), 13-14, 116, 203
    CTS-D, 35
    CTS-I, 14, 191
    defined, 204
Change Orders (CO), 43, 112, 134-135, 155
    defined, 203
    in checklists, 77, 148, 184-185, 188, 191
Charge-back, 113
Code, *see also software*
    building, 27, 37, 85
    National Electric, 204, 205
    of ethics, 14
    officials, 37

Commissioning, 46, 130, 154
  agents, 37, 61, 78, 156, 158
  and testing, 146-147, 160-163
  defined, 205
  in checklists, 77, 128, 149, 170, 180, 189, 200
  in contracts, 68-69, 111, 115
  report, 163
Communications, 175-177, *see also Data/telecom*
  and program report, 93
  contract, 113, 127
  Division 27-Communications, *see MasterFormat*
  Industries Association, *see InfoComm International*
  interfacing, 152
  of design intent, 110
  project, 80-81, 104, 132-135, 148-149, 188-189, 195
  services and providers, 36, 133, 146, 157-158, 162-163, 170
Conferencing, video and audio, 8, 196, 149,
Construction, 130-152, *see also Base building, Contractors*
  contract relationships, 44-54, 70
  deliverables, 121
  forms and checklists, 148, *see also the Appendices*
  industry, 2, 15, 18, 20-21, 24, 174
  installation team, 31-33
  kick-off meeting, 135, 148-149
  manager, *see Construction Manager*
  phase, 21, 31-37, 77, 130-152, 197
  process, 20-21, 36, 44-46, 121, 140, 202
  project meetings, 136
  schedule, 131, 107
Construction administration (CA), 68-70, 99, 202
Construction documents (CD or CDs), 22, 76, 108, 111, 203
Construction Manager (CM or Program Manager), 25, 34, 88, 98, 104, 124, 131, 135, 184, 203, 205
Construction Change Directive (CCD), 135, 184-185
Construction Specifications Institute (CSI), *see MasterFormat*
Consultant, independent AV design, ii, 10, 12, 29, 40-44, 46-47, 49, 56- 58. 70, 74, 204, 88, 132, 156, *see also AV: designer, Independent Consultants in Audiovisual Technologies (ICAT)*
Consultant/Integrator Team Approach, 52
Consultant-Led Design-Bid-Build, 20-21, 40, 44-47, 139
  and commissioning, 156
  contracts, 46, 68-69, 84
  method selection chart, 50
  provider selection, 57-60
Consultant-Led Design-Build, 49-50, 67, 69
Consultants, *see also Engineers*

  acoustical, 25, 30
  AV, *see Consultant, independent AV*
  data/telecom, 25, 30
  electrical, 25, 28, 29
  life safety, 25, 30
  lighting, 25, 30
  mechanical/HVAC, 25, 28-29
  move, 25
  other trade-specific, 25, 31, 95
  plumbing, 25, 28
  security, 25, 30
  structural, 25, 29
Consulting engineers, *see Engineers*
Contingencies, 92, 97, 163
Contract, *see also Warranties, Service agreements, Request for proposal, Software and under each delivery method*
  administrative requirements, 64, 111
  AV integration contract issues, 70
  best practice, 70
  cabling contracts, 72
  closeout, 71, 156, 159
  construction, 112
  consulting, 68, 76-77
  deliverables, 110-112, 116, 147
    design-build, 118-119
    matrix, 121
  design-bid-build, 68, 84
  design-build, 67, 84
  design-only, 69
  direct, 33, 46-54, 70-71, 74-75, 111
  documents, 22, 114, 118, 121, 133, 149, *see also Construction Documents*
  forms and checklists, 181-192
  front end section, 112
  checklist, 127
  forms, 184
  general conditions, 64, 112
  impact of value engineering, 110
  installation-only, 68
  internal to owner, 53
  misinterpretation of, 111
  prime, 27, 31
    defined, 205
  program-only, 68
  project directory, 85
  representative, 25, 27
  sign-off, 166
  subcontracting, 27, 31-32, 44-54, 119, 141
    double-sub, avoidance of, 70-71
    forms, 184-185
    markups, 97
    software, 73-75, 78
  submittals, 136
  template, 67, 114, 184-185
  terms and payments, 27, 64, 70, 85, 110
    deposit and mobilization fee, 113, 127
    payment terms, 113-114, 127
    penalty clause, 113, 127, 135
    progress payments, 113-114

Contracting, 56-78, *see also Contracts, General Contractor, Subcontractor*
Contractor, 25, 30, *see also General Contractor, Subcontractor, Contracts, and individual trades*
  AV Integrator, 32
  data/telecom, 32
  electrical, 32
  lighting, 25
  mechanical / HVAC, 25
  structural, 25
Control room, *see Master control room*
Control systems, 8, 66, 73-74, 81, 120, 122
  commissioning, 155-157, 162-163
  defined, 203, 205
  GUI / user interface, 33, 73-74, 120, 122, 124, 145, 162-163
    GUI in checklists, 100, 128, 193
  in checklists, 100, 102, 128, 149, 151, 170, 188, 193
  in contracts, 73-74, 133
  lighting interface, 30
  software programmers, 13, 33, 61, 74
  software programming, 33, 66, 74-75, 81, 120, 122, 145Cost cutting, 108, *see also Value engineering*
Cost estimate, 95, *see also Budget*
  breakdown, 96
  opinion of probable cost, 76-77, 95, 100, 106, 122
  preliminary system estimate, 95
  quote, 95
  value engineering 108-109
Cost Estimator, 98
CSI (Construction Specifications Institute), *see MasterFormat*
CTS, CTS-D, CTS-I *see Certified Technology Specialist*
Customer Furnished Equipment (CFE), *see Owner Furnished Equipment*
Custom operation guide, 139, 146, 171

Dashboard for Controls, 123, 202
Data/telecom, *see also Structured cabling*
  and AV cabling, 72
  and construction phase, 130, 132, 140
  and design phase, 104, 106-107
  and MasterFormat, 54
  and program report, 92, 95
  commissioning/testing, 161-162
  consultant, 30
  contractor, 32
  defined, 204
  in checklists, 76, 101, 126, 148-150, 188, 193
  interdependence with AV, 30, 132-133
Dealer, *see AV: dealer*
Deliverables, *see Contract: deliverables*

Delivery methods, 44-54
    and deliverables, 112, 121
    overview, 20-21
    scoping for, 66
    selection chart, 50
Design, 8, 20-21
    and owner responsibilities, 188
    certification, 13-14
    checklists, 126-128
    consultants, 12, 27-31
    drawings and specifications, 21, 35, 48, 64, 68, 76, 111, 118, 122, 124, 127
    GUI, see Control Systems
    package, 110-118
    phase, 21, 104-125
    review, 123-125
    schedule, 43-44
    team, 25, 27Design-bid-build, see Consultant-Led Design-Bid-Build
Design-build, 20-21, 29, 31, 44, 48, 52, see also Integrator-led design-build
    contracts, 67-68, 76, 99, 112, 185
    defined, 204
    deliverables, 118
    process, 119
    RFP, 99, 127
Design intent, 35, 110, 112, 128, 154, 169, 203
Design-only contract, 69
Design phase, 20-21
    conceptual, 20-21, 107
    construction documents, 20-22, 76, 108, 203
    design development, 20-21, 76, 107, 204
    kick-off meeting, 104
    schematic, 20-21, 76, 107
Designer, 35, see also AV: designer, Interior designer, Consultants
Developer, 25, 33
Digital signage, 11, 177, 194-195
Digital Signal Processing (DSP), 161, 171
    software programming, 73, 81, 143, 205
    testing, 144, 157, 170
Direct sales, 18
Display systems market, 2
Distributors, 12
Division 27, see MasterFormat
DSP, see Digital Signal Processing

E/A (Engineer/Architect), 27, 204
Electrical contractor (EC), 25, 32, 204
    in checklists, 77, 190
    relative project schedule, 140
    subcontracting to, 46, 50, 70
Electrical engineer or consultant, 25, 28-29, 107, 204
Electronic(s)
    documents, 138, 147, 184-185, 195
    field production (EFP), 195
    markerboard, 196

    network, see Data/telecom
    news gathering, (ENG), 195
    signage network (ESN), 177
    systems design, see AV systems and Design
End-user, 27, see also Owner
    AVBP, ii, iv
    defined, 204
    during construction phase, 133, 139, 149
    during design phase, 104, 106, 122-124
    expertise, 27, 43-44, 92, 156, 167
    markets, 177
    needs, 81, 82
    non-technical, 26
    owner team, 25
    program, 82, 85, 88-89, 98, 193-196
    real, 26
    technical, 26
Engineers, 27, 35, 186-188, 202-206, see also Consultants
    Professional (PE), 15
Engineering, see also Design
    credentials, 15
    drawings, 85
    program phase, 92
    value, see value engineering
    services 186
Estimates, see Cost estimates
Ethernet connectivity, 133, 149, 157, 162, 170, see also Control systems

Facility manager, 25-26, 88, 94
    defined, 204
Fast-track, 31, 56, 67, 93, 98, 99, 106, 111
    defined, 204
Field order, see Construction Change Directive (CCD)
Final checkout, see Commissioning
Front-end, see Contract: front-end
Functionality, see AV systems: functionality
Furniture, Fixtures and Equipment (FF&E), 42, 184
    defined, 204

General Contractor (GC), 31, 71, 98, 130
    defined, 204
    in checklists, 148
Graphical user interface (GUI), see Control systems: GUI

High voltage, 28
    defined, 204
Home theater, 9, see Residential
HVAC (heating, ventilating and air conditioning), 28, 29, 85, 92, 107, 108, 130, 131, 133, 143, 145, 162, see Mechanical
    defined, 204
    in checklists, 101, 126, 149, 150, 188, 189

ICAT, see Independent Consultants in Audiovisual Technologies
Implementation, 8, 18, 68, 120
Independent Consultant-Led Design-Bid-Build, 41, 44, 46
Independent Consultants in Audiovisual Technologies (ICAT), ii, 204
    Dashboard for Controls, 123
Independent design consultant, see Consultant, independent AV design
Independent manufacturer's representative, 12
Independent programmer, 13, 33, 74, 75, 123, see Control systems
InfoComm, ii, 43, 202, 204
Information technology (IT), 11, 205, see also Data/telecom
    department, 27
    interdependence with AV, 30, 132-133
    network, 26, 174
Infrastructure
    budgeting, 34, 122
    commissioning, 162
    coordination, 104, 131, 141, 197
    defined, 204
    design, 20-21, 44-45, 81
    drawings, 115
    establishing, 106
    in checklists, 76-77, 101 126-127, 148, 150 170
    process, 105
    program phase, 91-92, 95, 193, 195
In-house capabilities, 11, 31, 44, 47- 53, 111, 164,
    in checklists, 190, 194, 205
    defined, 204
    software programmer, 74
Innovations in AV, 5, 177
Inspections, 37
Installation, see also Construction
    AV systems, 143, 145
    conditions, 65, 117, 134, 142
    contract, 70, 111, 130, 155, 168
    funding AV, 42
    in checklists, 77, 78, 102, 148- 151, 188, 189
    infrastructure, 131-133, 141
    installer, 35
    meetings, 135-136
    process, 137
    punchlist, 158
    team, 25, 31
Installer, 35, 36, 111, 142, 204
Integrated systems, 4, 12
Integration, 10, 31, 44, 70, 131, 141-146, 157-163
Integrator-Led Design-Build, 20-21, 48-49, 99
    and commissioning, 156
    contracts, 48, 68-69, 84
    method selection chart, 50
    provider selection, 57-60

Intellectual property, 137, 143, 181, 182
Internal politics, 27
Interactive systems, 7, 8
Interconnections, 27, 115
Interface, see Control systems: GUI
Interfacing
   industry, 8
   life safety, 30, 194, 196
   lighting, 30, 92, 131, 142, 148
   network, 72
   other systems, 100, 150, 152, 170, 189
   to control system, 120
Interior design, 10, 27, 184
Interior designer, 25, 28
International Communications Industries
   Association, Inc.® (ICIA®), ii, 2, 7, 13-17, 43, 116, 202, 204, see also InfoComm
   certification for companies, see Certified Audiovisual Solutions Provider
   standard contract template, 67
International Organization for Standardization (ISO), 15
Internet, 7, 11, 30, 36, 72, 205
   auctions, 60
Internet Protocol (IP), 132, 133, 157
   in checklists, 100, 149, 151, 170, 171, 189
Interoperability, 3, 162, 163
ISDN, 30, 36, 133, 146, 157, 162, 195, 196, 203, 205
   in checklists, 100, 149, 170, 171, 189
ISO 9001:2000, 15
Isolated Ground, 28, 205
   in checklists, 148
IT, see information technology

Kick-off meeting
   Design, 104
   Construction, 135, 148-149

LAN (Local area network),
   see Ethernet connectivity
Life safety, 25-29
   consultant, 30
Lighting, 8, 10, 12-13, 25
   commissioning, 160, 162
   construction phase, 130-132, 142
   consultant, 30
   design, 28-29, 92, 95, 104, 107-109
   in checklists, 76, 100-102, 126, 148, 150, 170, 188-189, 196
   interface and control, 131, 163
   market, 2-3
Live events, 4, 165
Local area network (LAN),
   see Ethernet connectivity
Low voltage, 10, 28, 32, 70, 130, 148
   interfaces, 189
   defined, 205

Maintenance, 16-17, 165
   OFE issues, 53
   preventive, 77, 154, 166, 168, 199
Management team, 25, 33
   building management agency, 34
   construction manager, 34
   developer, 33
   move consultant, 25
Manufacturers, 12, 111
   control systems, 120
   documentation, 146, 151, 171
   specifications, 118
   training, 43
   warranty 105, 139, 167, 171, 199
Markup, 46, 71, 92, 97
Master antenna distribution system, 162, 194, see also Cable TV
Master Control Room, 72, 194-195
MasterFormat and CSI, 54, 112, 202
   and AV specifications, 115-118, 128
Mechanical / HVAC
   consultant / engineer, 25, 28, 205
   contractor, 25, 32
   coordination, 145, 149-150, 188-189
   defined, 204
   drawings, 85
   engineer, 7
   infrastructure, 92, 95, 101, 104, 106-108, 126, 130-131, 133, 143
Meetings, see Kick-off meeting, Construction
MEP (mechanical, electrical, and plumbing), 27, 207, see also Consultants, Contractors
Millwork, 32, 130-131, 134, 140, 143
   defined, 205
   in checklists, 76, 126-127, 148, 150-151, 170, 189
Mock-ups and prototypes, 144, 158
Mounting details, 108, 115, 128, 138, 151
Move consultant, 25

National Certification Commission, 14, 116
National Defense Education Act, 5
NAVA (National Audio-Visual Association), 5, 7, 202
Needs analysis, 10, 20, 45, 49, 53, 67, 68, 69, 80, see Program
   defined, 80, 205
   pyramid, 81
Network, see Ethernet connectivity
Non-technical, 26, 57, 67, 124, 154, 165

Occupancy, 20, 21, 26, 37, 46, 50, 64, 104, 105, 140
Occupational safety, 142
   in checklists, 150
OFE, see owner furnished equipment
OFE / Integrator installed, 40, 52
OFE / Owner installed, 40, 53
On-site, 130, 145, 146. see also Site
   in checklists, 148, 149, 150, 170, 188

   occupational safety, 142, 143
   response, 199
   service, 168, 199
   support, 17
One-line drawing, 138, 139, 160
   defined, 138
One-off projects, 42
Operations, 42, 169
   and maintenance, 17, 27, 42, 53, 186, 204
   in checklists, 102, 170
   operational procedures, 171
   operation guide, 139, 146, 147, 171
   operations manual, 85, 151
Opinion of Probable Cost, 95, 96, 97, 106, 122, see also Cost estimate
   defined, 95
   in checklists, 76, 77, 100
Optical systems, 138, 161
Optimization, 154, 155, 162, 163
OSHA, 143
   in checklists, 188
Outsourcing, 194
Owner, 20, 24, 44, 74, 75, 82, 85, 88, 98, 106, 124, 133, see Owner furnished equipment
   in checklists, 102, 149, 150, 156, 188
   installed, 53, 54
   politics, 90
   representative, 27, 31, 46, 57, 113
   responsibilities at installation, 141
   sign-off, 166
   staffing, 11, 26-27-43-45, 53-54, 70-71, 75
      commissioning/training, 158, 165
      design phase, 124
      in checklists, 194
      program phase, 87, 92
   team, 25, 26
Owner Furnished Equipment, 32, 44, 106, 133, 141, 169
   in checklists, 101, 149, 150, 188
   defined, 205

Patch bay, 138, 148, 152
Performance evaluation, 156, 157, 161, 162
   see Commissioning
Plasma displays, 2, 3, 5, 142, 177
   in sample warranty, 199
Portable Document Format (PDF), 147, 171
Post commissioning, 154
POTS (plain old telephone service), 36
PowerPoint, Microsoft, 195
   for GUI submittal, 145
   originally Presenter, 5
Pre-assembly and testing, 130, 143
Pre-commissioning Tests, 146
   see Commissioning
Presentation professional, 13
Preventive maintenance (PM), 77, 154
   in sample warranty, 199
Princeton Public Library, 176

Privacy and security, 143, 150
Pro-AV, professional AV, 2-18, 202, 205, *see AV*
Probable cost(s), 92, 95, 96, 97, 122
    in checklists, 76, 77, 100
    opinion of, 95, 106
Process for AV project
    assessing the, 41-42
    choosing, 54
    Consultant-Led Design-Build, 49
    Consultant/Integrator Team Design-Build, 52
    Design-Bid-Build, 46
    Design-Build, 48
    OFE/Integrator Installed, 52
    OFE/Owner Installed, 53
Procurement, 89, 41, 130, 140
    auction, 60
    Brooks Act, 187
    legal and policy issues, 52
    requirements, 42
Professional engineer (PE) license, 15, *see also Engineers*
Program Manager
    see Construction Manager (CM)
Program Report, 68, 81, 83, 89, 90, 95, 98
    already available, 84
    architectural, 85
    defined, 80
    distribution, 97
    end-user/technology manager sign-off, 98
    in checklists, 76
    objectives, 93
    "shopping", 98
    structure of, 94
    system description, basis of, 99
    writing, 92-97
Programmer (software), 81, 122, 139, 144
    contracting, 74, 75
    control system, 33
    Independent, 13, 123
    third party, 144, 162
Programming (architectural), 21, 45, 80, 90
    see Needs Analysis
    architectural, 29, 92
    contracting, 84
    integrator, by, 69
    meeting agenda, 89-90, 101
    meetings and attendees, 88-90
    process 94
    vs. software, 81
Project
    budget, 22, 42, 106
    collaborative process, 18
    construction, part of, 34
    description/scope/summary, 76, 116
    process, 20-21, 44, 46-54
    schedule, 45, 64, 131, 140
    sign-off, 166
    success, 10
    teams, 24, 56
    this book, iii-iv

Project manager (PM)
    defined, 35
Project meeting, 136, *see Kick-off meeting*
    status reports, 136
Project parameters in building project team, 56
Project phases, 10, 20-21
    commissioning and training, 22, 153
    construction, 22, 32, 129
    design, 21, 103
    program, 20, 79
Proposal, 59, 99
    design-build, 119
    integrator's, 67
    program report, and, 99
    Request for (RFP), 60, 63
Prosumer, 9
Punch list, 71, 150, 154, 155, 156
    defined, 158
    generating, 158
    in checklist, 170
    in commissioning report, 163
    requirement for sign-off, 166
Purchasing agent, 27

Qualifications, 14, 56-57, 116-117
    architect-engineer (Form SF330), 62
    request for qualifications (RFQ), 60-63, 112
    selection methods, 58-60
    in checklists, 78, 102, 127-128, 185-187, 202
Quality assurance 41, 139
Quality control, 15, 24, 67, 124
Quality of Service (QoS), 132
Quote, 67, 96, 160, 167
    defined, 95
    equipment, 81

Rack fabrication, 139
Registered Communications Distribution Designer (RCDD), 15
Renovation, 27, 33, 37, 41, 46, 104, 130
Rental and staging, ii, 12, 13
Request for Change (RFC), 134-135
Request for Information (RFI), 68, 134, 135
Request for Qualifications (RFQ), 58, 60, 112
    defined, 61
Request for Proposal (RFP), 59, 60-68
    design-build, 99
    non-technical, 67
    technical, 68
Residential AV market, 3, 9
Response time, 61, 65, 168
RFI, *see Request for Information*
RFP, *see Request for Proposal*
RFQ, *see Request for Qualifications*
Roles, 24, 85
    design team, 27-30
    installation team, 31, 33, 35
    owner, 26-27, 132

    project manager, 35
Rough-in, 131, 142, 197
    in checklists, 150, 170

Satellite distribution, 6
SAVVI (Sound, AudioVisual and Video Integrators Council), ii
Scalability, 16-17
Schedule, 20, 31, 35, 40, 41, 64, 104, 135, 158
    AV fitting in, 45
    change orders, 135
    delivery method based on, 49-50
    Gantt chart and WBS, 131
    illustrated, 45, 50, 140
    project, 41, 131
    site, 145
    training, 165
Security consultant, 25, 30
Selection strategy/method, 57
    fee-based, 59
    qualifications-based, 58
    two-envelope, 58
Selection team, 57
Service agreement, 166-169, 185
Service record, 168
SF330 form, 62-63
"Shop", 79
    the design documents, 49
    the installation, 48
    the program document, 98
Shop drawings, 115, 119, 137-139, 142, 143, 160
    defined, complete, 138
    final documentation, 146
    preparing, 137
    submittal, 51, 136-137
Shop testing, 144, 159
Shipment, 143, 144
Sign-off, 21, 83, 98, 166-167
    in checklists, 188
Signal flow, 138, 151
Sightlines, 29, 76, 107, 126
    illustrated, 108
Site (job), 31, 130, 131
    inspection and conditions, 65, 71, 134
    installation, 145-147
    preparing, 141-143
    security and safety, 114, 143
    warranty, 61
Software, 8, 13, 120, 143
    contracting for, 73-75
    license agreement, 182
    testing, 143
    programming, 74, 75, 81, 120
Sound reinforcement, 2
Space requirements, 28, 45, 85
Specifications, 10, 21, 67, 68, 111, 118, 154, 157
    changes 134
    control system, 120, 122
    CSI, 115

defined, 111, 115-116
in RFP, 128
Standards, 154
AV, 5, 117-118, 160-161, 174
Classroom Acoustics, ANSI S12.60, 86
control systems GUI, 73, see also
Dashboard for Controls
facility, 26
forms, 62-64, 133, 184
in checklists, 101, 128, 149, 193, 199
industry, 86, 143, 207
National Electric Code, 205
of excellence, 16 see AVSP
owner, 26-27, 32, 85, 106
Status report, 136
Streaming media, 2-3, 8, 175
Structural, 27, 95, 107-108, 131
consultant, 25, 29
engineer, 15, 35
mounting, 108, 132, 142
Structured cabling, 30, 72, 133, 161
in checklists, 148, 193, 195
Subcontract(or), 27, 42, 65, 71, 74, 107-108, 131, 142, 142
AV integrator, 71
defined, 32
payments, 46, 113
programmer, 61, 74, 144
Submittals, 21, 37, 104, 151
approval, 137
GUI, 145
preparing, 136
process for, 113, 115
requirements, 117
Subsystem deletions, 109
Substantial Completion, 71, 146
defined, 159
forms, 184-185, 198, 199
Report of, 159-160, 198
Sub-subcontract, 70-71
System commissioning, 51, 68, 146, 157-166
defined, 154
who and when, 156
Systems, 9, 85, 115
control, 120
design, 110
Integrators, 12
value engineering, 108

Tax-exempt, 73
Teams, 18, 20, 24, 56, 116
consultant and integrator, 52, 132
design, 27, 88, 93
install, 31, 98
management, 33
owner, 26
project, 56-57, 104, 124
Technical
consultants, 28
documentation, 85

end-user, 26, 124
funding, 43
RFP, 68
roles, 28, 35
staff, 27, 61,124, 193, 194
training, 165
Technical subcommittee, 27
Technology-enhanced, 18, 45
Technology manager, 25, 27, 88, 98, 102, 193, 203, see AV manager
Technology Managers/End-Users Council, 17
Dashboard for Controls, 123
Telecommunications, 8, 10, 72, 172
Test gear *(see also equipment)*, 161
Testing, 37, 50, 68, 118, 141, 142
during commissioning, 157, 159-170
for certification, 13, 14
preliminary, 146, 157
shop, 144
site, 146
subjective, 163
Third party
commissioning agent, 37, 156-162
programmer, 144
software license, 181
training, 43
Training, 12-13, 16-17, 7, 43, 77, 118
commissioning phase, 21, 147, 154, 164-168
defined, 205
in checklists, 128, 149, 150, 189, 194-195, 198
Transient Voltage and Surge Suppression systems (TVSS), 28, 148
Trinity Church, 176
Two-envelope bid, 58

Updates, software, 169
Upgrades, 41-42, 73, 94
in checklists, 102, 195
UPS (Uninterruptible power source), 28
Usage, 3, 41, 73, 86, 154, 168
User manuals, 137, 139

Value, 16-17
project award, 33, 57
Value engineering, 108-109
affects contracts, 110
Brooks Act, 186
cost cutting, 108-109
Verification, 84, 154, 157
site conditions, 188
Vertical market, 42, 49, 57, 86
RFQ checklist, 78
Video conferencing, 36, 133, 146, 165, 193
in checklist, 149
Video production, 13, 194, 195
Visual display, 8

Wadsworth, Raymond, iii
Warranty, 139, 159, 166-169, 198
in checklist, 77, 78, 127,170-171
period, 21, 22
response time, 61
sample statement, 199
sign-off and start, 166
software, 182
Webcasting, 2, 3, 11, 175
Web-enable, 3
Westchester Medical Center, 176
White boards, 2, 3, 175
Wireless, 3, 175, 177
connectivity, 2
in checklist, 170
testing, 157
user interface, 163
Work breakdown structure (WBS), 131, 197
Workmanship, 161
in checklist, 127, 128
warranty, 166-167

VIII. Authors' Biographies

Co-Authors

Timothy W. Cape, CTS-D
Principal Consultant
Technitect, LLC

Timothy Cape has been an independent consultant in the fields of audiovisual systems, distance education, videoconferencing, acoustics, lighting and related disciplines since 1982. He regularly teaches on these subjects both nationally and internationally. He is also an active AV industry advocate and a prolific writer. He currently writes the monthly column "Consultant Connection" for Pro AV magazine. After being an owner of two successful AV consulting firms, Tim launched his third consulting firm, Technitect, LLC in 2004.

He is a faculty member of InfoComm International's InfoComm Academy Audiovisual Design School that supports the CTS-D certification program, and is an active member of the consultants' councils for both InfoComm International and the NSCA. Mr. Cape holds a Bachelor of Science degree in Physics from the University of Washington in Seattle and has held the CTS-D (Certified Technical Specialist – Design) certification from InfoComm International since 2000. www.technitect.com

Michael J. Smith, Jr., CTS
Director of Strategic Initiatives
HB Communications, Inc., North Haven, CT

Jim Smith began his audiovisual career in 1972. He has worked in all parts of the industry as a technician, system designer, business owner, instructor, consultant, and writer. In 1980 he became media director for a large high-tech company in Cambridge, Massachusetts, where he was responsible for video production, facility design, and conference management.

HB Communications, a large AV integration company, recruited Jim in 1991 to manage sales and operations of the Boston region. In 2002, Jim became a member of HB's corporate management team as Director of Strategic Initiatives. In this position, he develops processes for improved quality, efficiency, and standardization. Special projects include exploration of new vertical markets, development of user interfaces, and writing for various trade publications.

Jim currently serves on the InfoComm International Board of Governors as Chairman of the Sound, Audio Visual, and Video Integrator's Council (SAVVI).